Surface and Apparition

Surface and Apparition

The Immateriality of Modern Surface

Edited by
Yeseung Lee

BLOOMSBURY VISUAL ARTS
LONDON • NEW YORK • OXFORD • NEW DELHI • SYDNEY

BLOOMSBURY VISUAL ARTS
Bloomsbury Publishing Plc
50 Bedford Square, London, WC1B 3DP, UK
1385 Broadway, New York, NY 10018, USA

BLOOMSBURY, BLOOMSBURY VISUAL ARTS and the Diana logo are
trademarks of Bloomsbury Publishing Plc

First published in Great Britain 2021

Cover design by Louise Dugdale
Cover image © Lesley Halliwell, *Shroud* (2016) Gold leaf, pencil, ballpoint pen and
lacquer on paper, 75 x 60 cm. Photography by Tony Richards. Copyright Lesley Halliwell.

A catalogue record for this book is available from the British Library.

Library of Congress Cataloging-in-Publication Data
Names: Lee, Yeseung, editor.
Title: Surface and apparition : the immateriality of modern surface / edited by Yeseung Lee.
Description: London ; New York : Bloomsbury Visual Arts, 2020. |
Includes bibliographical references and index.
Identifiers: LCCN 2020024151 (print) | LCCN 2020024152 (ebook) | ISBN 9781350130449
(HB) | ISBN 9781350130456 (ePub) | ISBN 9781350130463 (ePDF)
Subjects: LCSH: Surfaces (Technology)–Design. | Surfaces (Technology)–Psychological
aspects. | Texture (Art) | Decoration and ornament. | Aesthetics.
Classification: LCC TA418 .S87 2020 (print) | LCC TA418 (ebook) | DDC 620/.44–dc23
LC record available at https://lccn.loc.gov/2020024151
LC ebook record available at https://lccn.loc.gov/2020024152

ISBN: HB: 978-1-3501-3044-9
ePDF: 978-1-3501-3046-3
eBook: 978-1-3501-3045-6

Typeset by Integra Software Services Pvt. Ltd.
Printed and bound in Great Britain

Contents

List of Contributors

Jane Birkin is Research Fellow at Winchester School of Art, University of Southampton, where in 2015 she completed a practice-based PhD entitled 'Units of Description: Writing and Reading the "Archived" Photograph'. She also works on exhibitions in Archives and Manuscripts at the University of Southampton Library, as exhibition designer and maker as well as in a curatorial role. She is ever-mindful of the many debates around the digitization of archive objects. Taking the archive as the primary locus of her own practice, in particular the status of the archived image, Birkin unfolds the term 'archive' through film, text and lecture.

Giuliana Bruno is Emmet Blakeney Gleason Professor of Visual and Environmental Studies at Harvard University. She is internationally known for her research on the intersections of the visual arts, architecture, film and media. Her seminal book *Atlas of Emotion: Journeys in Art, Architecture, and Film* (2002) provided new directions for visual studies and won the 2004 Kraszna-Krausz Award for 'the world's best book on the moving image'. Other books include *Streetwalking on a Ruined Map* (1993), winner of the Society of Cinema and Media Studies book award, and *Public Intimacy: Architecture and the Visual Arts* (2007). Bruno has contributed to numerous monographs on contemporary visual artists, including Isaac Julien's *Riot* (2014) and Chantal Akerman's *Too Far, Too Close* (2012), and has written for the Whitney Museum's show *Dreamlands: Immersive Cinema and Art, 1905–2017.* In *Surface: Matters of Aesthetics, Materiality, and Media* (2014), she revisits the concept of materiality in contemporary visual culture. She is at work on a new book entitled *Atmospheres of Projection: Environments of Art and Screen Media*.

Benedict Carpenter van Barthold is Principal Lecturer in Fine Art at Nottingham Trent University. He was trained at Chelsea College of Art and Design and the Royal College of Art. He was awarded the Jerwood Sculpture Prize in 2001 and has completed a number of permanently sited sculptural commissions in the UK and Europe through Arts Council England, the EU initiative Leader + and others. Carpenter van Barthold has a persistent academic and creative interest in art medals as small sculptures to be appreciated in the hand. A number of his own medals are in public collections in Norway and the UK. Since 2012, his research has focused on tactility and the material negotiation of intent in art and craft practice.

Oksana Chefranova is Research Scholar in the Film and Media Studies Program at Yale University. She received a PhD in Cinema Studies from New York University, and her research bridges film, media and built environments such as gardens, fairgrounds, film settings, theatre stages and ruins. She is working on her monograph, *From Garden to Kino: Evgenii Bauer, Cinema, and Genealogies of Built Environment in Russia Circa 1900*. Recently, Chefranova contributed to 'Breathing Faces, Twinkling Eyes: On Cinematic Visage in Russian Films of the 1910s' in *Corporeality and Early Cinema* (2018). Her other projects focus on history and theory of camera movement, early film theory, landscape across media, experimental film and art practice, the artists' moving image and new approaches to film style and aesthetics.

Lesley Halliwell is undertaking a practice-based PhD (AHRC) at Manchester Metropolitan University (2014–20). She studied at Dartington College of Arts, trained as a painter at Nottingham Trent University and then went on to study Art History (Goldsmith's College, MA) and Fine Art (Manchester Metropolitan University, MA). Halliwell has exhibited in *Beauty is the First Test* (Pumphouse Gallery, London); *The Drawing Show* (Castlefield Gallery, Manchester); *Jerwood Drawing Prize*; *Superabundant* (Turner Contemporary, Margate); *Bloombergs New Contemporaries*; and *Pattern Recognition* (Leicester City Art Gallery). Recently, her work 'Pattern: Entry and Reflection' was published in *Drawing: Research, Theory, Practice* (2019).

Freyja Hartzell is Assistant Professor of Modern Design, Architecture and Art at the Bard Graduate Center in New York City. She received her PhD in the History of Art from Yale University and her MA in Decorative Arts, Design History and Material Culture from the Bard Graduate Center. Her research interests include the cultural politics of design; object studies (ontologies, agency, things); emotion studies, empathy, and theories of perception; and materiality and immateriality. She is currently completing her first book, *Living Things: Richard Riemerschmid and the Modern Object*, and is engaged in research for a second book on glass, transparency and modernity in the nineteenth and twentieth centuries.

Joanne Horton is Senior Lecturer in Mixed Media Textiles at De Montfort University, Leicester. Her doctoral research (2018) has been informed by a strategic search for new or underused ways to create decorative metal surfaces.

Elaine Igoe is Senior Lecturer in Textiles and Fashion at the University of Portsmouth and an Associate Lecturer in Textiles Research in the Department of Textiles at the Royal College of Art. Her research is focused on textile and material design theory. Her book *Textile Design Theory in the Making* is forthcoming in 2020.

Petra Tjitske Kalshoven has been a cultural anthropologist with the University of Manchester since 2009 (PhD McGill University, Montréal, 2006) and explores skilled manifestations of human curiosity and play. She is the author of *Crafting 'the Indian': Knowledge, Desire, and Play in Indianist Reenactment* (2012), an ethnographic study of the social and performative dynamics of a contemporary amateur practice in Europe predicated on expert replication of Native American life worlds from the past. More recently, her research has centred on the emulation of nature that underpins the skilled practice of taxidermy as a lens for thinking through human–animal relations and nature–culture dichotomies. Since 2017, as a member of the University of Manchester's social and nuclear sciences research network *The Beam*, Kalshoven has pursued her interest in expertise, materials and landscapes with an ethnography of nuclear decommissioning in West Cumbria.

Yeseung Lee is VC2020 Lecturer in Design Cultures at De Montfort University. She has widely practiced in the international fashion industry in both the high-end and mass-produced markets (1998–2017) and holds a PhD from the Royal College of Art (2013). Her research draws on material culture, critical theory and psychoanalysis as a way of interrogating contemporary fashion with a particular focus on its materiality and praxis. Lee is the author of *Seamlessness: Making and (Un)knowing in Fashion Practice* (2016) and has published papers on fashion, textiles, skin studies and luxury studies.

Ray Lucas is Reader in Architecture at the Manchester School of Architecture, Manchester Metropolitan University, where he is Director of Humanities. His work is situated across architecture and anthropology, producing 'graphic anthropologies' as mixed-method enquiries using various forms: drawing, notation, photography, filmmaking and mapping. The role of drawing as a form of knowledge production lies at the heart of his work since his PhD in Social Anthropology from the University of Aberdeen with the thesis 'Towards a Theory of Notation as a Thinking Tool'. Lucas is author of *Research Methods for Architecture* (2016), *Drawing Parallels: Knowledge Production in Axonometric, Isometric and Oblique Drawings* (2019), *Anthropology for Architects: Social Relations and the Built Environment* (2020), and is co-editor of *Architecture, Festival and the City* (2018).

Max Mosscrop is an independent artist. He studied architecture at the University of Liverpool and painting at the Royal Academy of Arts. He was awarded the NatWest Painting Prize in 1997 and has exhibited internationally. Since 2013, he has been developing a weaving workshop, building looms and making woven textiles. His recent work presents weaving as a sculptural adventure, in which the act, the equipment and the woven textile have equal significance.

Jussi Parikka is Professor at the Winchester School of Art, University of Southampton and visiting researcher at FAMU, Prague. His books include the media ecology-trilogy that consists of *Digital Contagions* (2007, 2nd edition

2016), *Insect Media* (2010) and *A Geology of Media* (2015), which addresses the environmental contexts of media culture and arts. He is known for his work in media archaeology including the key text *What is Media Archaeology* (2012) and he has edited various books such as *Writing and Unwriting (Media) Art History* (2015, with Joasia Krysa) on the Finnish media art pioneer Erkki Kurenniemi. He is also the co-editor of *Across and Beyond: A Transmediale Reader on Post-digital Practices, Concepts, and Institutions* (2016, with Ryan Bishop, Kristoffer Gansing and Elvia Wilk). Parikka's website/blog is at http://jussiparikka.net.

Barbara Rauch is Associate Professor at OCAD University, Toronto, Canada. She is a Digital Futures Initiative hire in the School of Interdisciplinary Studies and Graduate Studies. As an artist and research academic, Rauch investigates the affect and emotion in data analysis, using 3D printing and sculpting. Her studio aims to instigate discourse around the topics of data/information and materialization in artistic context. Her recent research focuses on developments in artificial intelligence (AI), machine learning and the suboptimal within the Anthropocene discussion.

Acknowledgements

The editor would first like to thank the contributing authors for their creativity, scholarship and professionalism. The collection presented here is the result of international collaboration by scholars and artists from the United Kingdom, the United States and Canada. I offer special thanks to the artists and designers whose original works have been reproduced within this book.

I would like to record my thanks to Professor Janis Jefferies, Professor Brooks Hagan, Fabrizio Poltronieri, Professor Sandy Isenstadt, Stephen Knott, Claire Pajaczkowska, Jane Norris and other anonymous peer reviewers who generously gave up a portion of their summer holidays to offer rigorous and constructive feedback for individual contributing authors. I also thank Bloomsbury and their five undogmatic anonymous reviewers who welcomed the concept I proposed and offered invaluable insights at the final stage of the manuscript preparation. I also wish to acknowledge the support I have received from my institution without which this book would not have been possible: Professor Siobhan Keenan, Professor Deborah Cartmell and Professor Kelley Wilder generously supported the project with VC2020 funding and Institute of Art and Design funding, and Helen Weeks and other colleagues at the ADH Research and Innovation Office patiently helped me through the paperwork.

This book is the final outcome of the interdisciplinary conference, 'Apparition: The (Im)materiality of Modern Surface' (The Great Hall of Leicester Castle, De Montfort University, March 2018), which I co-convened with Ellen Sampson. I am grateful to Ellen and colleagues of Fashion Research Network (Alexis Romano and Nathaniel Dafydd Beard) for their help with the conference. I am very grateful to Professor Peter Ford, Professor Carolyn Hardacker, David Heap, Jenny Gilbert, Elizabeth Lambourn and others in the Design Cultures team at DMU for their support with the preparation of the conference. I would also like to thank Diviesh Navekar, Thomas Hayes and Jasdeep Singh of DMU ITMS, Chris Sewart of ADH admin and the event assistants Tara Sood and Raabia Arif for their technical and operational support.

I thank Bloomsbury's helpful editors Claire Collins, Rebecca Barden and Olivia Davies. Special thanks go to Professor Jussi Parikka, whose encouragement and advice was indispensable in the process of preparing the book proposal, and Professor Giuliana Bruno, who replied to my unsolicited email and warmly embraced the project. I give my very special thanks to Professor Susanne Küchler and Professor Jonathan Faiers for their continuing support.

Prologue. Surface Matters

Giuliana Bruno

What is the place of materiality in our contemporary cultural atmosphere? To engage materiality, I suggest that we think about surfaces rather than images, and that we explore the fabrics of the visual, the surface tension of media and the changing atmospheres of projection. In order to pursue this new materialism, I propose performing critical acts of investigation on the surface, focusing especially on the translucency of screen surface, and mobilizing the wide potential of material expression across 'screens' of various media.[1] I have long argued for a shift in our focus away from the optic and towards the haptic, in order to understand the tangible spatiality of the visual arts, their moving, habitable sites, and the intimate experience they offer us as we walk through their public spaces (see Bruno 2002). The *haptic*, a relational mode derived from the sense of touch, is what makes us 'able to come into contact with' things. This reciprocal *contact* between us and objects or environments occurs on the surface. It is by way of such tangible, 'superficial' contact that we apprehend the art object and the space of art, turning contact into the communicative interface of a public intimacy.

This is why I prefer to speak of surfaces rather than images: to experience how the visual manifests itself materially on the surface of things, where time becomes material space. Digging into layers of imaging and threading through their surfaces, my theoretical interweaving emphasizes the material *fabrics* of the visual: the surface condition, the textural manifestation, and the support of a work as well as the way in which it is sited in space, whether on the canvas, the wall or the screen. I am particularly interested in what we may call the phenomenon of the 'becoming screen', that is, in the play of materiality that is brought together in light on different, intersecting screens, and in offering a theorization of the actual fabric of the screen of projection as a material surface-space. And finally, I am interested in the migratory patterns of such visual fabrications, and in tracing their material history as well as their shifting geographies. In this way, I want to rethink materiality and to show how surface matters as a place of intermediation and the transformation across mediums.

Surface matters in visual fabrics

In this age of virtuality, with its rapidly changing materials and media, what role can materiality have? How is it fashioned in the arts or manifested in technology? Could it be refashioned? I argue that materiality is not a question of materials but rather concerns the substance of material relations. Hence, in order to open a theoretical space for a reinvention of materiality, I want first to consider how the surface mediates material relations.

Surfaces are our skins. They are the epidermis of our bodies, the coat of paint on our walls, the façade of a building, the fabric of a painting's canvas, the plane for the projection of images. Surfaces, that is, are not superficial. They envelop us at all times in our lives. In this sense, surface is something temporal. Any surface is constantly exposed to the effects of time, the ambiance it lives within, and the environment that surrounds it. Surface also can be dressed, and even dressed up. This is an envelope that is sported and worn. But in wearing, a surface wears out. These surfaces that we call skin, fabric, canvas, wall and screen are all membranes that are positively 'consumed'. In the course of being worn they wear off. In other words, surface really wears. It is weathered.

This is because a surface is porous. It is a canvas that absorbs all that comes into contact with it. It takes in everything, all that it touches or that it is touched by, starting with the weather. Building façades are exposed to the elements, to meteorological conditions, and are changed by this contact over time. And this is also the case for our bodies. Our skin, our face, like the façade of a building, or the canvas of a painting, responds to atmospheric agents. It is changed by the surroundings. It is affected by rain, air, wind and sunshine. It responds to light and to changes of luminosity. In this way, surface is not only altered in time but shows the very course of time, even its memory. And so it wrinkles, warps and bends. It folds and furrows, crinkles and creases. It rumples and crumples.

I propose that we pay close attention to this haptic, ambient play of surface. In engaging the pleats and folds that constitute the fabric of the visual – its wear and weathering – I wish in this regard to expand on what Gilles Deleuze calls a 'texturology': a philosophical and aesthetic conception of art in which its 'matter is clothed, with "clothed" signifying … the very fabric or clothing, the texture enveloping' (Deleuze 1993: 115). To highlight this textural shift involves tracing what we might call the enveloping 'fashioning' of the surface, and weaving this across different media.

This requires first of all thinking of images materially, for it means viewing them as textures, traces and even stains. The visual text is fundamentally textural, and in many different ways. Its form has real substance. It is made out of layers and tissues. It contains strata, sediments and deposits. It is constituted as an imprint, which always leaves behind a trace. A visual text is also textural for the ways in which it can show the patterns of history, in the form of a coating, a 'film', or a stain. One can say that a visual text can really *wear* its own history, inscribed as an imprint onto its textural surface. It can also show affects in this way. After all, the motion of an emotion can itself be drafted onto the surface, in the shape of a line or a figure, and in the haptic thickness of pigment, and this atmosphere can be tracked down with tracking shots. An affect is actually 'worn' on the surface as it is threaded through time in the form of the residual stains, traces and textures that are felt in the air. In other words, surfaces not only bear the trace of time passing but show the mark of their own temporality in space. This is precisely the remnant, ambient quality of surfaces, the kind of wear that is weathered. Because surfaces create atmosphere, in all senses of the word, they affect us. In visual culture surface matters, and it has depth.

To understand materiality, we thus need to expose the work of the surface and show how textural matters manifest themselves in time and space. As we plumb the depth of surfaces that surround visual culture and see how they envelop us, we can observe how this porous material not only makes but also changes an atmosphere, carrying on the vestiges of time and communicating its residues in future reinvention. In other words, we can observe forms of transformation. Skimming the surface, we can not only weave together the filaments of visual existence, exposing their traces in layers of experience, but also trace patterns of transmutation. Surface especially matters as a site in which different forms of mediation, transfer and transformation can take place.

The surface tension of media: screen, canvas, wall

If we focus on surface as such a tensile, permeable, connective element, prone to respond to and also communicate atmosphere, we can see it actually acting as an agent of transformation. An important material transformation occurs as images travel across the surface of different media. Many changes affected by the migration of images happen on the surface and manifest themselves texturally in the form of a kind of surface tension, which affects the very 'skin' of images and the space of their circulation. In this sense, I claim that aesthetic encounters are actually 'mediated' on the surface and that such mediated encounters engage forms of projection, transmission and transmutation.

This surface tension has resurfaced today as an important manifestation of the culture of our times. It has emerged as a central condition of contemporary visual art and architecture, and driven an exchange of forms. To explore how art, architecture, fashion, design, film and the body all share deep engagement with superficial matters, in fact, means to show that surfaces act as connective threads between art forms as they structure our communicative existence. In 'superficial' forms of transit between media, we can see processes of contamination and hybridity taking place.

In many contemporary practices, this superficial awareness is not an incidental part of a work but is rather pushed to the limit of its potential to become the actual core and structure of the piece. I am particularly interested in this structural engagement with surface when it signals a refashioning of elemental agency and atmospheric materiality on our cultural screens. Sometimes, in fact, the matter of surface is exposed as a form of connection with the surface of the earth or becomes a way to enhance the skin of things, the depth of atmospheric sensations or the variety of meteorological layers. This aesthetic process that makes ornament into structure forces us to engage deeply with surface materiality, and to see how it operates across different media in ambient terms.

I will offer some examples to make this aspect of a surface tension more concrete in relation to the question of atmospherics that I am raising. In contemporary architecture, as the work of Herzog & de Meuron exemplifies, the façades of buildings are engaged as surfaces of ambient transformation. Lighter and more tensile, these surfaces may be energized by luminous play, texturally decorated as if they were canvas, stretched as membranes and treated increasingly as envelopes. Surface is here a responsive material that shows its own wear, exhibiting an organic process of meteorological ruination and connecting to other vital materials that wear out (see, among others, Leatherbarrow and Mostafavi 2002; Ursprung 2002). In contemporary art as well, such surface tension has emerged as a textural form of refashioning visual space across different mediums and, as a concept, is driving an aesthetic development that emphasizes an atmospheric redressing of space, attentive to the transformation of mediatic atmospheres (see, for example, Joselit 2006; Iles 2007; Coblentz 2003). Such ambient wearing of surface is an important phenomenon that art and architecture also share with cinema.

Think of the moody cinema of Wong Kar-wai, in which atmospheric forms of imaging are stitched together on the surface in patterns of visual tailoring. Here is a dense, floating surface in which one senses the material of light and the fabric of colour, emphasized by the visual pleating of editing that itself creates volume and depth, grain and granularity, the final effect being that residue and sedimentation appear retained in the saturated surface. We almost never see clearly through the fabric of this screen, for several coatings and planar surfaces are constructed out of different materials, and all are folded together. There are so many layers to traverse on the surface that the screen itself, layered like cloth, takes on volume and becomes a space of real dimension.

When a surface condition is activated in this way on visual planes, it changes our notion of what constitutes the support of the image and its way of siting a medium. I want to suggest that this new form of materialism initiates a major transformation, which can be called a form of projection. If we closely consider this textural process of fashioning the image, the very nature of what we have traditionally understood as canvas and wall changes to incorporate another form: the screen. An architecture of mediatic transformations comes to the surface at this very junction. Surface tension can turn both façade and framed picture into something resembling a screen. This filmic screen, far from representing any perspectival ideal, is no longer a window or a mirror but is reconfigured as a different surface. Made of translucent fabric, this screen is closer to a canvas, a sheet or a curtain. Partition, shelter and veil, it can be a permeable architectural envelope. On this material level, the current intersection of canvas, wall and screen is a site in which distinctions between inside and outside temporally dissolve into the depth of surface. The screen itself is becoming a translucent intermedial platform, in which all appears to fold back into screen surface – that reflective, fibrous canvas texturally dressed by luminous projections.

The connective, wearable surface of projection, which resonates with ambiance and dialogues with the environment, can indeed enable deep passage and transfer. As a membrane that mediates between subjects and produces communication with objects, surface not only creates but transforms the very fabric of projection. Projection is, after all, itself a form of transfer. In this sense, the translucent surfaces of the screens that surround us today function as our faces and façades, to convey the virtual transformation of our material relations. These screens, which have become new membranes of contact, exist in our environments in ever closer relation to the surfaces of canvas and walls, which are themselves porous cultural screens. And so it is here – in this meeting place that is the surface – that art forms are becoming reconnected and creating new hybrid forms of admixture and transformation.

An archaeology of migrant media

This 'becoming screen' is thus a fundamental aspect of our contemporary visual culture. Such a phenomenon is at the centre of a material reconfiguration of the very atmosphere of projection. It also should be noted that the translucent screen has become an ever-present, daily material condition of viewing, contact, and interaction, just at the point that cinema, at the very moment of film's own obsolescence, has come to inhabit today's museums. An intensification of this phenomenon is now taking place in a proliferation and exchange of screens of projection and communication. Such refashioning of atmospheres of projection shows tension at the edges, in the space beyond a medium, in the interstices between forms of art and experience, at junctures where both transgressive and transitive moves between the arts become palpable on the surface.

The screen acts as the actual surface of this refashioning by reinventing the absorptive materiality of a permeable space of contact and projection. As screen-based art practices enact such a return to materiality by emphasizing atmospheres of surface luminosity and ambiences of textural hapticity, the memory of film can also come to be materialized in contemporary art. As epitomized in Christian Marclay's video timepiece *The Clock* (2010), the history of film is learned in the art gallery. The screen is furthermore activated outside of cinema as a historically dense space – re-enacted, that is, as a mnemonic canvas that is fundamentally linked to the technology of light and its forms of projection. Walking through the art gallery and the museum, we encounter webs of cinematic situations, reimagined as if collected together and recollected on a translucent screen that is now a wall, a partition, a veil or even a curtain. The screen is therefore itself transformed as a material through these passages of light, activated in shades of translucency. And doing so, different enveloping fabrics of projection can come to light in this kind of surface tension, in dialogue with history.

This is also because the tensile surface of the screen canvas contains several 'sheets' of the past, which, unfolded, lead all the way back to the birth of modern vision and its history of visual surfaces. After all, the public museum flourished in the same age as the cinema, and shares with film that fabrication which is the visual, theatrical architecture of spectatorship. In a way, then, some of today's artists appear to be winking at pre-cinema, at the very phantasmagoric atmosphere out of which cinema historically emerged as a visual medium in public space. Artists are becoming archivists. They are turning into materialist scholars. Why? What is to be learned from this material history of surfaces? Can we refashion it for the future? If museum culture and forms of projection are mined as an archive of visual fabrics open to reinvention, this cultural archaeology of media can reveal the potential for artistic media to serve as the material conditions for haptic screen encounters.

I am interested in exploring this cultural history of screening when media archaeology does not involve simple nostalgia for, or celebration of, obsolete mediums of projection, dusting them up for spectacular effects or revamping them for digital delight. When artists pursue a conceptual and aesthetic exploration of the act of projection, the surface tension of media can be pushed in new, performative ways. If understood artistically as a potential space of transformation, the history of the surface of projection can in fact offer avenues to express what was not previously expressed, including unrealized potentialities that were left out of the canon of film history and traditional art history, thus projecting future possibilities of reception and material relations in projection.

The atmosphere of projective surfaces

This atmosphere of material transformations is the main object of my work in visual and spatial studies. It is especially important to engage the virtual movements that are taking place in material ways in this luminous environment of surfaces of projection, travelling across mediums. This passage is crucial because it affects the sedimentation of the visual imaginary, its residues and transformations. It concerns not the medium per se but rather the space of image circulation, forms of siting, and the situational experience that constitutes a visual environment. This is why I insist on the depth of surface, because we are speaking not simply of a medium but rather a perceptual field of mediations, an area of connections, a site of hybridity and mixture – that is, of an actual *ambiance*.

The exchange that takes place in such 'superficial' intermedial space, on the field screen of visual archives, profoundly affects the fabric and architecture of the visual experience, transforming its very atmosphere. In suggesting that we weave through the surfaces of material relations that link together screen, canvas and wall across time, exposing the threads that connect the visual to the spatial arts, including the migrations of forms of projection between art forms, my aim is not only to reclaim materiality but also to foster further explorations in the transformation of cultural paradigms. For the future of a medium presents itself texturally and experientially on the surface – that is to say, in the folds of its architecture, the thickness of its visual-culture history, and the tensility of its moving geography, affecting material change in cultural atmospheres.

Note

1 This text expands upon a central concern of my book *Surface: Matters of Aesthetics, Materiality, and Media*. It also addresses research interests that propel my new book, *Atmospheres of Projection: Environments of Art and Screen Media*.

Introduction: Surface, Haecceities, the Topology of Ice Deserts

Yeseung Lee

Reflecting on the conditions of art's emergence, Elizabeth Grosz (2008: 4) suggests its fundamentally relational character:

> Art is the regulation and organization of its materials – paint, canvas, concrete, steel, marble, words, sounds, bodily movements, indeed *any* materials – according to self-imposed constraints, the creation of forms through which these materials come to generate and intensify sensation and thus directly impact living bodies, organs, nervous systems.

Quite unlike human-centred approaches to art that function under the regime of signs or representations, art here is understood as a self-organizing process through which various materials, each already conveying haecceities,[1] selectively surface in a way that produces affects or intensities. Art, understood as such, may then encompass any type of production, insofar as it generates impacting forces and changes (ibid.: 3). Grosz is guided by Gilles Deleuze and Félix Guattari (1987), who interrogate the emergence of forms in general, beyond the arts. Their ideas of the machine – the machinic phylum, the machinic assemblage, the war machine, abstract machine, and so on – conceives the continuous surfacing of biological, technological and cultural forms in terms of 'matter-flow' and its immanent ability to assemble and reproduce itself. As humans are simply a part of such machinic processes, this matter-flow can only be followed (Deleuze and Guattari 1987: 409). It is this attention to movement, change and the line of differentiation that is also the focus of this volume. If emergence and becoming occur in the process of relating, we shall discuss surfaces rather than forms, as it is in surfaces that haecceities meet and draw a creative line of flight (ibid.: 422).[2] By following the im/material surfaces of everyday and artistic environments, this interdisciplinary volume explores how artistic or technological innovation surfaces from such matter-flow.

If the Cartesian–Newtonian understanding of matter as predictable substances supports human's conceptual and practical domination of the material world, new scientific understanding of matter as elusive and complex calls for critiques of ontological dualism (Coole and Frost 2010: 8–9): this involves rethinking human as only a part of the immanent vitality of matter and its continuous individuations and becomings. Therefore, in approaching the surface with a non-hylomorphic attitude, our subject matter is not the human intentions that surfaces may contain, or meanings that may be inscribed on the surface. Instead, it is the elective coming together of materials, implements, humans, desires and affects in the here and now, and how new ways of this 'coming together' occur: the process of surfacing and the resulting surfaces. By mapping and following the process of surfacing – 'the selective action of the assemblages upon the phylum' (Deleuze and Guattari 1987: 407) – this volume reveals the type of knowledge that can only be produced in the intermezzi,[3] in the mingling of making and thinking from various disciplines.

As we increasingly experience the world as surfaces, the changing technologies that design or manufacture these surfaces, in turn, make us. In contrast to the responses to preceding industrial revolutions, contemporary concerns with surface seem preoccupied with its function of mediation or passage, rather than with that of separation or boundary. Interestingly this dissolution of surface often occurs in tandem with insistence on its materiality. In this volume, scholars and practitioners in the fields of design, architecture, film, media, fine art, archival practices and anthropology account for how the material and the immaterial draw attention to each other, focusing on the changing technologies and processes that make everyday and artistic surfaces. Existing surfaces are scored, cut, carved, grooved, travelled on, polluted, layered or become porous, while new surfaces, for better or worse, inevitably emerge. Because surfaces are always continuously being made, we approach the surface in terms of its change and its ability to change anything that comes in contact with it; the way extrinsic relations – rhizomatic alliances or alloys, instead of arborescent filiations (ibid.: 25) – are made and unmade, how surfacing is always interfacing, or rather, poly-facing.

Following the process of surfacing means, then, following various heterogeneous components of matter-flow and instances of their various dis/assemblage. Therefore, each chapter, written by a practitioner or a scholar sensitive to current practice in their respective fields, addresses particular materials (cloth, paper, silver, copper, wood, resin, sandstone, nuclear waste, database, light or air), technologies (human sensory organs, manually operated tools or machines, industrial machines, digital devices and robots), modes of attention and movement, and affects. They account for the instances of material practice as 'events', as repeated processes of deterritorialization and reterritorialization, capable of drawing a new plane of consistency (ibid.: 422–3).[4]

This introduction first sets out the contribution this volume makes to the emerging 'surface scholarship' by positioning it within existing literature. It then considers the theoretical and practical issues that the book throws up, exploring why an attention to surfaces is vital today. In particular, I take relational and processual approaches to surface and surfacing, via Deleuze and Guattari's machinism; 'surface ontology' (Küchler 2008); and semantic devices that estrange common dualistic assumptions regarding the surface. The final section explains how the contributions come together as an assemblage via surface 'method' and introduces each chapter, pointing out some of the most salient lines running across the chapters.

Surfacing affinities

Surface is one of the most intensely debated topics in recent scholarship in arts, humanities and social sciences. The significant attention paid to various kinds of surface is reflected in proliferating publications and conferences held in fields as diverse as sociology (e.g. Adkins and Lury 2009), human geography (e.g. Forsyth et al. 2013), anthropology (e.g. Anusas and Simonetti 2020), design history,[5] early modern history,[6] or architecture (e.g. Hedges, Engels-Schwarzpaul and Jenner 2017). Many interdisciplinary exhibitions have also been held on the theme, accompanied by a catalogue weaving together the works exhibited and critical texts. To position this volume within these lively activities and their outputs, a few examples which have close thematic or organizational affinities with this volume are brought together below.

In a 1992 publication, *Incorporations*, the modern body and its macro- and micro-systems are seen in the relation of mutual incorporation. The essays and photographs by interdisciplinary artists, scientists and scholars illustrate how, throughout the twentieth century, overlapping 'biotechnic' arrangements have transformed a 'lifeworld' (Crary and Kwinter 1992). Almost thirty years on, we explore the increasingly uncertain organic–inorganic boundaries, especially how, in the process of making, humans, tools, information, social and cultural milieu can collectively function as 'autopoietic'[7] machines. More recently, in *The Virtual Window: From Alberti to Microsoft* (2006), film and media theorist Anne Friedberg's ultimate concern is the dissolution of screen surface and human sensorium into a virtual plane. As computer screens, televisions, glass walls, architectural windows, paintings and virtual reality technologies in public and domestic spaces gradually merge, in the future 'the screen may dissolve; images and data will be "uploaded" directly, bypassing the eye and the optics of vision' (Friedberg 2006: 244). Radical as this view may sound, with the growing dematerialization in the history of image – 'a territory-wall-painting-window-mirror-screen-becoming' (Grosz 2008: 17) – the current and future location of the human–non-human sensory interface is contingent and unpredictable. Both exhibitions, *Sensorium: Embodied Experience, Technology, and Contemporary Art* (MIT Visual Arts Center, 2006) and *Sk-Interfaces* (FACT Liverpool, 2008), and their catalogues (Jones 2006; Hauser 2008) explore the techno–human interface in order to rethink contemporary sensory apparatus as much more interconnected than ocular-centric Western culture would have it. In both publications, art, science, philosophy and technology relate with each other through surfaces. This volume similarly shows how creative practices and their interdisciplinary critical appraisals can destabilize existing understanding of what it means to be human in a changing technological environment. The idea of the body – of human, of technical machines, of im/materials – and its contentious territories are repeatedly brought into question via focusing on the process of their surfacing.

In the context of human geography, the special issue of *Environment and Planning A*, 'What are Surfaces?' (2013), rethinks surfaces as multiple, embodied and practised material productions. The editors suggest that we re-conceptualize the surface to challenge conventional ontologies and epistemologies by adopting relational approaches (especially non-representational theories) (Forsyth et al. 2013: 1015–16). A need to reformulate the link between depth and the superficial is a consistent thread in publications on the theme of surface that have recently proliferated. In this regard, *Surface Tensions* (2013) benefits from the methods and viewpoints of design history, by contrasting the historical preoccupation with the surface as superficial with contemporary efforts to complicate its understanding (Adamson and Kelley 2013: 2). The surface here is conceived as 'a site where complex forces meet', with the 'forces' referring mostly to people and things meeting via the acts of patterning, finishing and maintaining (ibid.: 1). In the special edition of *Theory, Culture & Society*, 'Visualizing Surfaces, Surfacing Vision', critical and practice-based contributions collectively emphasize the essentially processual quality of surfaces. The editors highlight 'how a focus on surfaces is at the same time a focus on (its) surfacing' (Coleman and Oakley-Brown 2017: 7), as 'a surface is always in the process of – and is constituted through – its surfacing: a surface is its becoming' (ibid.: 8). This emphasis on surface-becoming is what this volume also insists on, approaching it with more multi- and inter-sensorial modes of engagement within the process of material-making.

This rhizomatic review of recent works brings us to Giuliana Bruno's prologue to this volume, 'Surface Matters' and her monograph *Surface: Matters of Aesthetics, Materiality, and Media* (2014). In *Surface*, Bruno's (2014: 2) main concern is how the space of material relations manifest themselves on the surface of different media. Similar to Lyotard's assertion on page 11, in the age of virtuality, relations are made more explicit and material. As the contact between humans and our environment occurs on the surface, it is the surface – 'a quality of "becoming" as a connective, pervasive, or enveloping substance' – that allows us to engage with material relations (Bruno 2014: 3, 5). Surfaces and atmospheres enter a relational mode through touch or touch-like mediation, and it is also through this surfacing touch that the book *Surface* becomes interdisciplinary, bringing together art, architecture, fashion, design, film, new media and the body. This volume builds on Bruno's works and other existing literature touched on above, with a strong emphasis on practice and process, extending across disciplines and theoretical approaches. In particular, the four practice-based, visual-led contributions convey the complexity of surface beyond the textual, evincing that surfaces *are* relations; surfacing is the process of relating.

Surface, the machine and a clock of variable speeds

Writing about the conditions of British working life in industrial capitalism, socialist historian E. P. Thompson (1967: 57) asks: 'When the watch is worn about the neck it lies in proximity to the less regular beating of the heart. … how far did it influence the inward apprehension of time of working people?' The wider diffusion of the clock during the Industrial Revolution evidences that technological innovations subtly influence the way we are, as it irreversibly changed human conception and perception of time, habits and labour-discipline. Skilled clockmakers played a critically important role in the early stages of the Industrial Revolution, inventing and constructing much of the machinery used in the textile factories (ibid.: 65). The many technical innovations of the Industrial Revolution further intensified 'technological conditioning' (ibid.: 80), the ever-closer encounter between the mechanical and the organic. The French inventor Jacques de Vaucanson, whose mechanical loom was the basis of the automatic Jacquard loom, also designed a silk factory near Lyon that is often considered to be the first modern industrial plant (Foster 1993: 131). Vaucanson upheld the Cartesian mechanist view of the human body – distinct from the immaterial mind or soul – believing its workings to be analogous to the controlled movement of clockwork. His 'flute player' (1737), the famous automaton using clockwork mechanisms, epitomizes the material (body)–immaterial (mind) dualism in early modern Europe, which continued through the Industrial Revolution to the twentieth-century ideals of Taylorism.[8] As advanced technology allows more sophisticated ways of measuring and monitoring human resources, it could be argued that the now-ubiquitous 'digital Taylorism' brings the mechanical and the organic even closer. Coming true, then, is Donna Haraway's (1991: 150) prophesy: 'Modern production seems like a dream of cyborg colonization work, a dream that makes the nightmare of Taylorism seem idyllic.'

In the course of ongoing technological 'revolutions', works being carried out in fields such as biotechnology, artificial intelligence and neuroscience play a pivotal role in unsettling the Western transcendental and humanist subject. Artistic and scholarly interests today lie less in the rational, free and self-aware human mind presiding over material stuff, and instead lie in how the human is a part of an interconnected system which is also comprised of non-humans. A long line of theories and ideas that rethink the relationship between humans and technology is a good case in point. Anthropologist Marcel Mauss's 'Techniques of the body' ([1935] 2006), for example, touches on how the ways we use the body are conditioned by proto-machines such as shoes, hands or benches. Philosopher Georges Canguilhem (1992: 45, 55), in his 1947 lecture 'Machine et organisme', takes an anti-Cartesian non-dualist stance: putting forward the notion of 'organology', he suggests the use of certain biological principles to understand the construction of the machine. By contrast, cybernetic perspective, pioneered by mathematician Norbert Weiner (1948), regards living systems as machines working via the principle of closed feedback loop – with no material specificity. Inspired by cybernetics, neurophysiologist Manfred E. Clynes and psychologist Nathan S. Kline (1960) introduced the term 'cyborg', for unconsciously auto-regulating man-machine systems. The aim of such systems was to adapt human bodily functions to their extraterrestrial environment during long-term space voyages by incorporating exogenous devices (Clynes and Kline 1960: 27, 76).[9] The idea is resolutely Cartesian dualist, as it imagines the human mind capable of thinking and exploring, as if unaffected by the body integrated within the human-machine feedback loop (Hacking 1998: 209–10). Such an ontological separation between human and machine is rigorously rejected in Bernard Stiegler's (1998) idea of 'technogenesis' which suggests human's co-evolution with their technological environment, and Andy Clark (2003) similarly argues that we have always been 'natural-born cyborgs' or 'human-technology symbionts'.

Out of various works and ideas exploring the inseparable link between humans and technology, it is through Deleuze and Guattari's idea of the machine that I want to provide a rationale for this volume to collectively propose. In many of their collaborative or respective works, Deleuze and Guattari suggest the idea of the machine as 'the functional ensemble' including various components such as human, material, energy, plans, information, etc. (Guattari [1992] 1995: 34), as the technical object is determined by its extrinsic relations or an 'ecosystem' (Deleuze and Guattari 1987: 395). Whether an object is a weapon or a tool is thus 'nothing but consequences' (ibid.: 398). In this context, the term 'machine' refers to a 'synthesis of heterogeneities' (ibid.: 330): when heterogeneous elements enter into combination with each other 'through recurrence and communications' (Deleuze and Guattari 2009: 91–2), they constitute a machine or a 'machinic assemblage'. Such machine, still a 'multiplicity of distinct elements' (ibid.: 112), is *unpredictable*, *non-teleological* and capable of producing 'events'. On that account, the relationship between machines and organisms is also explained in terms of their in/ability to produce *differences*. The organism ('the organization of organs') is regular, static, binary and territorializing, whereas the machine is 'a body without organs'[10] composed of nomadic singularities and intensities, constantly drawing the line of flight (Deleuze and Guattari 1987: 40–1). For Deleuze and Guattari (ibid.: 150–1), the need to find your body without organs is 'a question of life and death', because 'nothing happens anymore' in an organized and stratified world: you must find out how to 'walk on your head, sing with your sinuses, see through your skin, breathe with your belly …' – find out how to become the machine.

They call attention to paleoanthropologist André Leroi-Gourhan's study in *L'Homme et la matière* (1943), in which he takes biological evolution in general as the model for technical evolution. Leroi-Gourhan's thesis is based on the fact that the technical object, as the living being, is developed within conditions established by the milieu (Stiegler 1998: 45–6). Accordingly, Félix Guattari ([1992] 1995: 33) extends the biological idea of 'autopoiesis' (Maturana and Varela 1980) to technical, cognitive, affective and social domains to explain the machinic assemblage as the process of innovation or a movement of individuation. Autopoiesis refers to self-organizing aspects of living systems as they continually self-repair in relation to the changes happening in their external environment. Therefore, unlike a cybernetic structure with insular feedback loops, an autopoietic machine would be ontologically dependent on heterogeneous exterior elements (Guattari [1992] 1995: 37) and is capable of becoming a new assemblage possessing 'phylogenetic differences' (ibid.: 42). By adopting the biological terms 'phylum' or 'phylogenesis' in describing the changes of non-organic forms, Deleuze and Guattari suggest ontological continuity between the organic and the mechanical as they co-mutate.

In the process of assembling, disassembling and reassembling, a 'war machine' (a new phylum or a new shoot of a rhizome) emerges: an industrial innovation, a technological invention, a scientific or artistic movement can all be a war machine, insofar as they draw, in relation to a phylum, a creative line of flight, fostering further new connections (Deleuze and Guattari 1987: 360, 422–3). A machinic assemblage, in this sense, is a veritable invention and genuine creation (ibid.: 406). We must, therefore, conceive the modification of cultural forms in terms of capabilities immanent in the matter-flow, and assemblages surfacing in changing configurations. This volume maps various ways of following the matter-flow, as each contribution enquires into the selective action of the assemblages upon the phylum (ibid.: 407). Matter has powerful surfacing capacities of its own, and the surface gives access to how such surfacing occurs.

Surface ontology

Writing between 1920 and 1931, cultural critic Siegfried Kracauer (1995: 75) analyses the modern condition, brought on by technology and mechanized production, from 'surface-level expressions'. With the thriving urban mass and extended nightlife, surfaces of popular culture (picture palaces, circuses, theatres) and commercial display (arcades, advertising, hotel lobbies) were filled with 'surface splendor' (ibid.: 323). Modernity's obsession with the surface was played out through built and manufactured everyday material things. Factory spaces full of gleaming machinery were worshipped like a temple; the sleek surface of Bakelite signalled a new era of affordable consumer goods; Josephine Baker wore her naked skin like a shimmering sheath (Cheng 2011), an idealized surface of industrial capitalism. The link between the changing cultural environment and various surfaces in the everyday and artistic contexts was also discussed in many of Walter Benjamin's essays. Benjamin ([1939] 2003: 251) begins the third and final version of his essay 'The Work of Art in the Age of its Technological Reproducibility' with Paul Valéry's prescient comment, written in 1928: 'the amazing growth of our techniques, [and] the ideas and habits they are creating, make it a certainty that profound changes are impending in the ancient craft of the Beautiful'. Valéry goes on to suggest that technological innovations will transform the entire techniques of the arts and even our very notion of what art is. Indeed, with new material and technological possibilities, we attain new ways of sensing – more distributed, shared with or enhanced by computation, in or outside of the body – and subsequently, of perceiving and knowing. For example, Anne Friedberg (2006: 194) argues that, with the computer screen on which multiple 'windows' overlap, our new mode of perception is more multiple and fractured than it was in pre-PC eras. Since 2007, properties of various media such as film, TV and computer are brought together in smart touchscreens, consistently 'interfacing' with humans. Beyond glass screens, computational technology is being discretely embedded in everyday surfaces, including the human body. These surfaces are 'intelligent', 'smart' and networked: they respond to, record and watch our heart rates, touch, voice, habits, eye movement and sleeping patterns. It is not surprising then, that popular culture and creative fields display a renewed preoccupation with the surface. If previous eras were preoccupied with stable surfaces as the limits, now surfaces are being reimagined as vapour, air, light, vibration, perfume or powder, acquiring ephemerality. They dissolve, fizz and effloresce. In many popular ASMR videos,[11] surface interactions dominate the scene with amplified sounds of grating, crushing, cutting through, scratching or stroking all kinds of surfaces, which may suggest the heightened responsiveness of visual-auditory-haptic sensorium. These surfaces do not signify; they instead enter into composition with the viewer-listener to become a sort of assemblage that combines affects and intensities. Many artists and designers challenge existing notions of surface, operating between computational and non-computational realms, thereby producing new types of art and products. Such post-digital modes of working are ubiquitous in contemporary art, design and architecture. David Berry (2015: 52) defines the post-digital as a relation produced by various surfaces embedded with computation, which is hegemonic, entangled with every aspect of human experience. This would mean that aesthetics and the everyday must always be reflected on in relation to the computational (ibid.).

Therefore, the renewed preoccupation with the surface or its dissolution – rather than being a reflection of 'depthlessness' within the cultural logic of late capitalism, as Frederic Jameson (1990: 99) regarded seemingly weightless architectural surfaces – is brought on by the changing technology outlined above, and by our heightened awareness of the fluid and distributed modes of being that this change affords. Surface is our becoming. Criticizing the negative use of the term 'superficial', Daniel Miller (2005: 53) points out that 'depth ontology' is pervasive in Western philosophy, 'whereby we tend to assume that everything that is important for our sense of being lies in some deep interior'. Surface in such contexts is a boundary, delimiting relations with the 'outside'. For that reason, anthropologist Susanne Küchler (2008: 116) suggests the need for a 'surface ontology', as smart, sensing and networked membrane-like surfaces now create at once a material and informational environment. Our concern, then, is not looking under the surface, but instead, looking *at* surfaces in constant action and modification; or rather, looking between surfaces as a 'spatial, inter-artifactual modality' (ibid.: 104). As these surfaces function in close proximity with the human body and its quotidian life, reacting to various affective responses, it is now essential that we bring our interaction with them into the notion of social relations. As the idea of the autopoietic machine makes clear, this state of 'transhuman' has always been the case, but becomes more explicit in computerized, sensorized and networked environments today. Surface ontology, or surface-becoming, is a multiple and relational mode of being, in which human and non-human, material and immaterial, surface and space are radically continuous. Surface needs to be rediscovered over and over, if we acknowledge Deleuze and Guattari's (1987: 257) assertion:

> We know nothing about a body until we know what it can do, in other words, what its affects are, how they can or cannot enter into composition with other affects, with the affects of another body, either to destroy that body or to be destroyed by it, either to exchange actions and passions with it or to join with it in composing a more powerful body.

We can now reimagine surface as a flexible and im/material dimension that expands and contracts along the flow of relating. Surface is thus a practical achievement, that is, something that must always be created, as it is not a site or container of relations; it is rather the very body of relations.

Outplaying[12] dualism: floating devices

As briefly touched on at the outset, Deleuze and Guattari use the idea of machine to do away with the hylomorphic model and its 'organic chauvinism' (DeLanda 1997). The hylomorphic model assumes the genesis of form as imposed from the *outside* (typically God or humans) of supposedly inert matter. Negating transcendental agency hidden in the background, Deleuze and Guattari emphasize that forms are immanent to matter itself.[13] It is haecceities, or degrees of intensity, that bring about individuals or events by assembling themselves and forming a machine in turn (Deleuze and Guattari 1986: 86). Biological, geological and social systems are all spontaneously self-generated from haecceities. The hylomorphic model is also based on binary oppositions – matter and form at each extreme of a teleological development – ignoring the existence of intermediary dimension. In Deleuze and Guattari's ontology, nothing progresses or develops; instead, 'things arrive late or in advance, and enter into some assemblage according to their compositions of speed. [... things] grow from the middle, to be always-in-between' (Deleuze and Parnet 2007: 93). Like the Schumannian body in *Kreisleriana* (Opus 16), machinic assemblage keeps diverging: the body *stretches*, extends to become a corpuscular space (Deleuze and Guattari 1987: 273).

Critical of depth ontology and hylomorphism, each chapter of this volume interrogates dualism or binary assumptions via surfacing operations: layering, growing, grooving, fluting, weaving, mixing, folding, oscillating ... all reveal the intermediary dimensions that emerge as surfaces surface. This collective attitude is reflected in the word 'immaterial' in the title of this book. The term conveys a binary relationship between two things by implicitly containing an apparition-like stroke ('im/material'), because the most common use of this sign is to denote alternatives, as in either/or, on/off, s/he (Waddingham 2014: 90). What Deleuze and Guattari (1987: 20–1) try to achieve in *A Thousand Plateaus* is precisely to 'invoke one dualism only in order to challenge another', thereby to 'arrive at the magic formula we all seek – PLURALISM = MONISM – via all the dualisms that are [the] entirely necessary enemy, the furniture we are forever rearranging'. Similarly, in a series of lectures entitled *The Neutral*, Roland Barthes (2005) suggests that we annul the binarism of A/B via 'a back-and-forth, an amoral oscillation' (Barthes 1997: 132). Rather than defying boundaries by merging or being neither A nor B, Barthes chooses to float or waver while exploiting existing binaries. Such movements let us experience the intermediary area – the space of 'and ... and ... and ...' (Deleuze and Guattari 1987: 25) – which is composed of differences, haecceities and nuances, rather than abstract generalities or averages. In an attempt to avoid the inherently oppositional structure of meaning and discourse in Western culture, Barthes (2005: 51, 211), as a literary semiologist, deploys a grammatical genre 'the neutral' to register the nuanced space in-between, replete with slight differences. The 'immaterial', therefore, outplays dualistic paradigms by highlighting existing assumptions and choosing, instead, to multiply, float and waver. Via the operation of the neutral, it hopes to do what Jean-François Lyotard tried to achieve with the title of his exhibition *Les Immatériaux* (Centre Pompidou, 1985): to show how the human-centred conception of the material is altered by new materials (Hui and Broeckmann 2015: 10) and how relations are made explicit and *material* under digital conditions (Hui 2015: 131). 'Material' and 'immaterial' are not considered in opposition to each other, allowing a floating space for multiplicities.

The desire for the neutral, the desire to create a nuance that 'skips the paradigm' (Barthes 2005: 51) is also present in the term 'apparition' in the title of this volume. As a figure of indistinction, apparitions defy binary oppositions such as presence and absence, body and spirit, past and present, life and death (Buse and Stott 1999: 10). In this book, this figure is variously deployed to explore mediatized fashion, invisible vitality of nuclear power, algorithmic creativity, the virtual augmented by the artist-machine combination, and so on. Surface-making here, as the operation of the neutral and apparition, dodges dualism by exploiting existing binaries: art and science, material and immaterial, virtual and actual, living and non-living, human and non-human. What emerges out of the surfacing operations performed by each of the contributions is that the practice of surfacing allows us to be more aware of relational modes of being, and how the process of relating is also a movement of differentiating. This would mean, conversely, that any study concerning changes and the line of change must engage with surface and its becoming.

Surface method

In this book, the surface functions as an interdisciplinary 'method' for attending to critical issues concerning creative and technological innovations. As a machinic assemblage, the book itself is a multiplicity of distinct elements, and manifolds[14] of extrinsic relations (interdisciplinary alliances) rather than of intrinsic relations (disciplinary filiations). The chapters grow in many directions, rhizomatically along the surface, with no chronological order or disciplinary boundaries, so that theories and practices that are not often discussed in a single volume come into contact with each other and become contagious. Akin to the topology of ice deserts or sand deserts that relies on haecceities – on sets of relations between 'winds, undulations of snow or sand, the song of the sand or the creaking of ice' (Deleuze and Guattari 1987: 382) – its unity is that of 'contagions, epidemics, the wind' (Deleuze and Parnet 2007: 69) circulating through surface-contacts. Multiple threads run across the chapters in non-linear ways. New assemblages may thus surface, potentially opening up spaces for 'methodologies without boundaries' (Denzin and Lincoln 2018: 8). The process of relating is the practice of surfacing. As surfaces emerge whenever and wherever relations occur, it is not possible to explore the theme in a comprehensive way. Therefore, this volume focuses on making surfaces and surfacing changes: each chapter is a critically engaged discussion written either by a practitioner on their own practice; by a practitioner on others' practice in their respective discipline; or by a scholar sensitive to current practice.

The main body of the book begins with Jussi Parikka's sharp political investigation into the planetary surface of the Anthropocene, exposing the smooth surface of global capitalism as an illusion generated by mediatized culture. Through the architecture studio Unknown Fields' video work *Unravelled*, which maps postcolonial aspects of the 'planetary conveyor belt' (globalized fashion production that relies on subcontracting and outsourcing), this chapter positions the book within the interdisciplinary fields that Giuliana Bruno brought together in *Surface*, while expanding the spatio-temporal scale into continent-crossing infrastructures and the Anthropocene. Surfaces here operate as points of tension between the visible and the invisible, and between micro- and macro-scales of movement, connecting the cloth upon our skin, the planetary surface, polluted rivers in foreign lands and the fingertips of exhausted workers at industrial sewing machines. Fashion is ephemeral yet material, and this is precisely the problem. The surface of fashion here exposes the environmental chains linking the living and the non-living, highlighting the urgency of conceiving the world as a web of interrelated processes, of seeing how the everyday choices we make as individuals have consequences for the world.

The planetary surface is also the main focus of Petra Tjitske Kalshoven's chapter, 'Surface-making in Nuclear Decommissioning'. Kalshoven narrates her anthropological fieldwork on the practice and discourse of nuclear waste storage in the UK, linking geologic time with the present. The continuity between living and non-living, which Deleuze and Guattari (1987: 411) explain via the example of 'panmetalism', is supported by complex technologies involved in creating artificial boundaries between human and radionuclides. Whether radioactive waste is stored on the surface of the earth or deep underground, the unknowns and uncertainties of nuclear waste are a constant reminder of 'thing-power' (Bennett 2010), forgotten in the context of waste disposal in general, which often involves movement across the planet, to be made into someone else's problem.

The living and non-living divide is complexified by the fascinating hybrid creatures borne out of, and evolving in, the digital realm: in Chapter 3, 'Surface Eruption: Machine Creativity and Emotive Data Objects', designer Barbara Rauch discusses the surfaces of algorithm-generated 'art'. As Rauch traces the process of 'data surfacing' – from AI-generated objects on computer screens to 2D- or 3D-printed physical objects – involving no physical human touch, she questions the origin of the distinct aesthetic surface qualities. Artefacts created by machines are now often indistinguishable from those created by humans.[15] As intelligent surfaces gather, analyse and learn our unconscious behaviours and habits, machine creations tend to reveal the human characteristics and cultures that produced them.[16] Can algorithms possess emotions, and ultimately, artistic creativity? God the creator, human the creator, and machine the creator merge in Rauch's creatures, which cut across multiple boundaries.

In compelling contrast, artist Lesley Halliwell's entirely handmade drawings in 'The Depth of Surface' look as if they are computer-generated. The drawings are created by painstakingly grooving patterns on paper and silver foil with a plastic Spirograph toy and biros. As Halliwell's methodical movement simultaneously coalesces (the rhythm of pattern and patterning) and erodes (material wear and oxidization, hesitations, imperfections) the surface, the artist's empathic relation to her medium creates contingencies (Latin *con-*, 'together' + *tangére*, 'to touch') (Doane 2012: 349), eventually opening an 'entry point' through which image turns into a kind of n-dimensional surface.[17] Silver is a material of visual illusion: early motion pictures, modern 3D films, our reflections in the mirror or in photography all rely on the material for their apparition. While human labour is usually invested in bringing out the reflective shine of silver, Halliwell's labour is invested to interrupt the shine in her conscious effort to float between seeing and touching, the virtual and the actual.

In Chapter 5, 'Where Surface Meets Depth', designer Elaine Igoe also approaches intersensorial transmutations through the devices of material swatch and computer screen. As a small tear on the surface of silver foil in Halliwell's *Portal* (2000; 2017) opens an n-dimensional surface, Igoe's swatch operates as a neutral passageway through which designers can oscillate between material and immaterial, presence and absence, optic and haptic. Through case studies of four emerging design studios that practise within the post-digital, Igoe bears out that virtual textiles and materials can deliver affect. As we also see in Mosscrop's fine art practice in Chapter 8, it is our relationship with the surface – our immersion in matter-flow and social machines – that generates intensities, converting the digitality of the textile to something analogue and affective.

In 'Growing Surface between Textiles and Electrochemistry', textile designer Joanne Horton is also a 'metal horticulturist'. Her copper-crystal embroideries are not sewn, but instead grown in a chemical solution, adopting the Victorian technology of electrometallurgy. This process of interfacing the living and the non-living, cloth and metal, growing and making, art and science brings about a resourceful attitude towards making, drawing a line of flight. Horton's repurposed and retrofitted tools lead us to the following three makers' creative use of tools, explicitly surfacing maker-tool-material-environment machines.

In Chapter 7, sculptor Benedict Carpenter van Barthold explores the fluted surface of a rosewood dish as an index of David Pye's uniquely mechanized system of production. In this system, the material (wood), the maker (Pye), the implement (the fluting engine, a hand-operated milling machine Pye invented) and their movement come together as a single machine. As Carpenter van Barthold traces the iterations of the fluting engine over a period of four decades, he is also able to investigate how the machinic ensemble changed. This combination of synchronic and diachronic analyses allows us to reflect on the nature of practice mediated by tools, the link between actual and virtual touch, and how the surface materializes the technology that made it. As his study shows, when the maker constructs or improvises their own implements, the elective character of assemblage becomes pronounced.

This aspect of the machinic phylum is also evidenced by artist Max Mosscrop's work, *Journal (2016–2018)*, discussed in Chapter 8. Mosscrop is a painter who weaves: his attempt to escape the binary opposition between figure and support in fine art painting led him to move from the painting frame, firstly to a wicker board as deconstructed frame, and then to weaving, which in turn led him to build his own wooden loom. The documented journey, or journal, reveals how the looms, the cloth and the weaver are in a relationship of co-production. Once set in motion, self-built, hand-operated machines like the fluting engine and Mosscrop's loom, seem to become almost indistinguishable from the maker's body, perhaps like a ballet dancer's customized pointe shoes. The resulting wood dish or cloth, then, is a thoroughly in/organic surface.

Architect and anthropologist Ray Lucas's drawings in Chapter 9 are also the surface of the in/organic assemblage: in this case, the lifeworld at Namdaemun Market, Seoul. Lucas's in- or post-situ inscriptive practices reproduce profuse surfaces of the market beyond their visual manifestations, conveying the unique relational system in which people and environments make each another. The drawings show how the politics of space in the market – with its people and ephemeral architecture of modular and mobile carts – is articulated through adaptive creation and use of surfaces. Lucas's method of grasping the vast and unknowable market through walking and inscribing is particularly salient in comparison with, and also in contrast to, disembodied internet searches: we can only 'surf', as any attempt at comprehensiveness would inevitably fail. This chapter also allows us to consider how and what we can know by artistically reproducing surfaces. The drawings of Leonardo da Vinci reveal his desire to understand the workings of nature and technology, and Lucas gains unique access to the lifeworld of the market through his drawings. The same question prompts Birkin's material translations through scanning and printing in Chapter 10.

In 'Archive Surface', Jane Birkin tackles 'depth ontology' head-on, by juxtaposing the archive document as linguistic information (depth) and material information (sensorial and affective surface). An exceptionally decomposed, tightly bound bundle of letters from the Wellington archive is ironically rendered an icon of literacy, precisely because the written information is inaccessible due to decay. Birkin intensifies this tension between textual and textural, between depth and surface, by 3D scanning and printing the bundle, thereby creating superficial surfaces in the computer screen and in the hollow 3D print. This surfacing adds another layer of tension: between original and copy. As Walter Benjamin ([1939] 2003) suggests, the process of copying reveals the mechanisms of the aura and how its authenticity is performed. Birkin successfully defamiliarizes the implicit dualism regarding language (depth, mind, cerebral) and material (superficial, body, peripheral) through the use of digital technology.

Historian Freyja Hartzell also performs an operation of the neutral in Chapter 11, 'Experience, Poverty, Transparency: The Modern Surface of Interwar Glass', by oscillating between the concept of the glass surface as an object for 'looking *at*' and for 'looking *through*'. As Hartzell traces glass objects and architecture produced in early twentieth-century Germany, the material moves from a surface appreciated for its material properties to a self-effacing transparent vessel for shifting human values and intentions. Taking advantage of the historian's point of view, Hartzell effectively demonstrates how attending to surface is attending to the renewed relationship between humans and non-human materials.

Richard Sennett (1992: 108) suggests that the use of transparent thermal plate glass in modern buildings, combined with air conditioning, isolated other senses from the optic, contributing to the dematerialization of everyday experience. In the final chapter, 'On Genealogy of the Translucent Screen and Rehabilitation of the Ephemeral', film theorist Oksana Chefranova explores how this type of dematerialization is countered by means of translucent screens in contemporary surroundings. With examples ranging from the nineteenth-century Diorama and spirit photography to contemporary post-cinema, installation art and performance, Chefranova suggests that the phenomenon of omnipresent translucent surfaces supports a post-digital aspiration to 'overcome the superficiality of the digital image'. The quality of translucency, a thoroughly nuanced surface spatiality, is another operation of the neutral. This surface, almost-there, floats and wavers between visible and invisible, figure and background, stage and screen, collapsing these binaries as a result. Chefranova approaches such apparition-like quality with a near meteorological awareness, as the 'screenness' here virtually touches our skin, like breeze or mist. These types of intersensorial transpositions open more questions about various interfaces and intermezzi.

The surface is a multiplicity, manifold, constantly surfacing. As our surfacing operations render surface even more complex than before, here would be a good place to address the element of uncertainty contemporary technological surfaces tend to generate, as computation becomes increasingly opaque and unknowable. To do this, I return to the figure of apparition. If the phantasmagoria of the eighteenth and nineteenth centuries effectively evoked Gothic apparitions that may appear anywhere at any time, our WiFi-enabled technological prostheses evoke contemporary apparitions, the 'uncanny of virtual locality' (Punter 2007: 133). With well-regulated surfaces (selfies and other curated personal information) featured on various social media platforms and wirelessly distributed around the globe, the contemporary ghost is truly omnipresent. Technology–human interface increasingly occurs on the level of emotions, beliefs and biases, but we do not have control over our personal data being put together as 'data doubles' to be manipulated by surveillance regimes (Haggerty and Ericson 2000). Furthermore, the techno–human interface is also creating 'ghost workers': the invisible labour force behind digital transactions often claimed to be powered by AI. Rather than replacing humans in low-paid menial tasks with machines, digital Taylorism conceals them from view. Amazon Mechanical Turk (MTurk), for instance, is a crowdsourcing site for employers to post tasks that computers are currently unable to do, such as moderating social media content (O'Brien 2019). E. P. Thompson's clock and its technical conditioning has come a long way. The figure of apparition, then, brings home to us that we are still far from grasping the ideal combination between human and machine.

These uncertainties hint that new technologies and globalization do not necessarily enhance our understanding of the world as a whole. This world cannot be comprehended by 'getting to the bottom of things', 'digging deeper' or 'peeling off the surface of things'. Surface ontology is helpful in overcoming the ideal of comprehensive knowledge and instead facing partial, contingent and topological views. This is because we can only follow particular combinations of human and non-human matter-flow as they surface in continuous variations. Through the process of surfacing, we 'get a sense of the irreducible social complexity characterizing the contemporary world' (DeLanda 2012: 42). If our future well-being depends on the extent to which we can nurture the generative processes of relating, we must pay attention to how surfaces are made, used, reused, repurposed and disposed. As Giuliana Bruno (2014: 8) suggests, rethinking materiality involves a refashioning of our contact with the environment to foster new forms of relatedness, and it is the surface – 'an architecture of relations' – that affords spaces for us to perform such transformations.

Notes

1 The term 'haecceity' is used in the medieval philosophy of Duns Scotus in order to designate the individuation of beings. Deleuze and Guattari adopt it in a unique sense to convey the process of individuation as an 'event', that is, *differentiation* (Deleuze and Parnet 2007: 169–70).

2 In the phrase *ligne de fuite* (line of flight), the French term *fuite*, translated as 'flight', conveys the sense of escaping rather than flying. For Deleuze, art functions as a line of flight by deterritorializing existing ways of being and thinking, thus generating new percepts and affects as a result (Parr 2005).

3 Roland Barthes (1991: 299) perceives the Schumannian body in *Kreisleriana* (Opus 16) as composed entirely of multiple intermezzi, with no organizing structure.

Inspired by this sensibility, Deleuze and Guattari (1987: 277) write: 'The only way to get outside the dualisms is to be-between, to pass between, the intermezzo.'

4 The plane of consistency has nothing to do with an interiority, memories, phantasms, the evolution or development of forms (Deleuze and Parnet 2007: 96–7). Instead, it is 'a plane of proliferation, peopling, contagion', in which 'form is constantly being dissolved, freeing times and speeds' (Deleuze and Guattari 1987: 267). For more on the plane of consistency in relation to the plane of transcendence, see the segment 'Memories of a Plan(e) Maker' in Deleuze and Guattari (1987: 265–71).

5 For example, Nic Maffei and Tom Fisher (2015) 'Unstable Surfaces: Slippery Meanings of Shiny Things' [conference paper], *Skin of Objects* (Norwich Castle Museum, 27 June).

6 See, for example, *Global Skins in Early Modern Europe, 1400–1700* (conference held at King's College London, 19–20 September 2019). https://renaissanceskin.ac.uk/events/conference-global-skins/ (accessed 1 October 2019).

7 For the notion of 'autopoietic' machines, see page 7 of this text.

8 Taylorism is the theory and practice of scientific management developed in the late nineteenth century by American engineer Frederick W. Taylor (1856–1915) to increase work efficiency and productivity. It aims to evaluate every step in a manufacturing process by breaking down production into specialized routines; by eliminating non-essential motions; by timing the workers; and linking pay to performance.

9 Clynes and Kline (1960: 74) offer the example of a 220-gramme rat – 'one of the first Cyborgs' – which has under its skin an osmotic pressure pump capsule continuously injecting biochemically active substances.

10 The body without organs is 'opposed not to the organs but to that organization of the organs called the organism' (Deleuze and Guattari 1987: 158).

11 The acronym for so-called 'Autonomous Sensory Meridian Response', referring to a sensation of euphoria triggered by certain types of audio prompts (Young 2019).

12 Barthes (2005: 6) explains that the neutral is 'that which outplays the paradigm', or 'everything that baffles the paradigm'. 'Outplay' is Rosalind Krauss and Denis Hollier's translation of the French word *déjouer*.

13 Manuel DeLanda (1998) explains how intensity differences, or haecceities, are morphogenetic: 'If one creates a container separated into two compartments, and one fills one compartment with cold air and the other with hot air, one thereby creates a system embodying a difference in intensity, the intensity in this case being temperature. If one then opens a small hole in the wall dividing the compartments, the intensity difference causes the onset of a spontaneous flow of air from one side to the other.'

14 The French term Deleuze uses in *Difference and Repetition* (1994) is the mathematical term *multiplicité* which translates as 'manifolds' in English (Patton 1994: xii).

15 In 2018, an AI-produced portrait entitled *Edmond de Belamy* was sold for US$432,500 at a Christie's auction in New York. It was created from a data set of 15,000 portraits painted between the fourteenth and the twentieth centuries, printed on canvas (Lawrie 2019).

16 In March 2016, Microsoft's Twitter chatbot, Tay, turned into a racist and sexist troll within twenty-four hours of its release. The system, in the process of learning the properties of language, also learned their historical cultural associations and acquired human-like biases in turn (Johnston 2017).

17 'A term in physics and mathematics, "n-dimensions" refers to an unspecified number of dimensions beyond our familiar three spatial dimensions and one temporal dimension. There are theoretical models that posit dimensions into the double digits' (Sobchack 2016: 299).

Chapter 1
Folds of Fashion: *Unravelled* and the Planetary Surface

Jussi Parikka

Figure 1.1 *Unravelled* (film still), *c.* 2017, directed by Unknown Fields in collaboration with Tushar Prakash.

Factory-floor fashion

A slow walk across the factory floor, a fashion model stylishly and gracefully gliding across the space with such elegance that one wonders whether she is a ghostly apparition or actually inhabiting the same high-ceilinged factory rooms as the workers busily immersed in their routines.[1] The simple golden dress, reminiscent of a sari, is one striking point of attraction as the garment moves in a row of images of movement, accompanied across three other screens by intense colours of red, blue and more. The walk across the factory floor is ceremonial, and while imitating the movement down the catwalk at a fashion show, it shows something that pertains more closely to the work and travels of fashion than to its mediated end results as fashion products. It shows multiple layers of materials, production, spaces and facilities as part of a condensation of fashion fabrics that is established by cinematic means. The improvised catwalk-style movement – a scene that occupies the whole of the video's seven-minute length, and is always on at least one screen of the four video channels – ends at a beach and a shipyard with massive, abandoned, rusting cargo ships. Also, this last scene seems to be out of place but indicating something essential about the off-screen of fashion's visual culture that is part of the usual repertoire of how it is seen: the work, logistics, infrastructures of fashion as media, fashion as film, fashion as movement that entwine with a multitude of other forms of aesthetics of modernity, including that of travelling across geographies (Bruno 2002). Hence, it concerns not only travelling as leisurely pleasure but the transport of mass-produced goods. One starts to ask: what are the forms of travel and crossing geographical surfaces that pertain to the cinematics and movement of fashion?

The experimental design studio Unknown Fields' recent video work *Unravelled* was exhibited in the early half of 2018 in the 'After the End of the World' show in Barcelona at the Centre de Cultura Contemporània de Barcelona.[2] The piece was part of a group exhibition that focused on the Anthropocene in theoretical and artistic contexts. It also continued the studio's earlier work on planetary scales of design culture. However, the four-screen video installation homed in on a different context of audiovisual culture from the earlier work, namely fashion film, in an extended sense that displays an understanding of the format that moves beyond mere branding and marketing: *Unravelled* approaches planetary logistics through garments, including some of the postcolonial contexts of their

Figure 1.2 *Unravelled* (film still), *c.* 2017, directed by Unknown Fields in collaboration with Tushar Prakash.

production. Hence, the beautiful, slowly unfolding video installation about fabrics and garments can be seen as an example of experimental practices that relate to the genre of fashion film, and of how contemporary design has extended its focus from objects to infrastructure, from things to their distributed nature.[3]

Unravelled is produced outside the fashion industry; it does not feature any particular brands, and Unknown Fields' previous work has focused primarily on the global logistics of urbanization in relation to methodologies of speculative design. Perhaps this earlier context is also one explanation as to how the installation engages with fashion in ways that do not feature the usual documentary language (cf. Rees-Roberts 2017) but become expressed in aesthetic shots that refer to conventions of the fashion show as well as to the extended sites where fashion takes place.

The video emerged from a 2017 expedition led by Liam Young and Kate Davies, described as a mix of landscape studies that mapped the rural conditions of contemporary media-cultural urbanization, including the fashion spectacle:

> For our 2017 Expedition we pick at a loose thread on the garment we are wearing and unravel it across continents from wardrobe to warehouse, from factory to field, in search of the landscapes behind the runway dreams and street blue jeans. Before we wear them, our clothes make journeys of tens of thousands of miles in their process of production making textiles the most globalized industry on the planet. The garment trade has for a long time played a critical role in the evolution of developing nations. As labour costs rise in China we follow the threads and travel through the factory floors and fabric mills now manufacturing in India and Bangladesh. Here iconic rivers run with the colours of the season as chemicals used in the dye process are dumped untreated to poison the land along the rainbow banks that will mark our trail from mountains to the sea, as we embark on a waterborne journey down river from India to the Ganges Delta in Bangladesh. We will visit the 'T-shirt cities' and 'textile valleys' that span from field to factory and will pick from vast cotton crops and silk-worm cocoons and draw yarns across deafening shuttles as rows and rows of automated looms weave the fashion fads of a distant world. Plain t-shirts, haute couture and throwaway high street chic all begin their lives in these landscapes.[4]

The film is shot in multiple locations including Chittagong (Bangladesh), Varanasi, Ahmedabad, Gurgaon and Pali (India), shifting views from textile to fabric dyeing, cotton fields to yarn production and garments; the locations provide the fabric around which the themes of materials and production are cinematically composed together.

I want to consider how this video installation, as images of fabric, sits as part of recent discussions about fashion, moving images and also fashion film.

Figure 1.3 *Unravelled* (film still), *c.* 2017, directed by Unknown Fields in collaboration with Tushar Prakash.

Recent work such as Nick Rees-Roberts's (2017) has made a convincing case as to how some documentary-style productions can be also considered part of the extended visual culture of fashion film as it spreads across multiple platforms, not least online. Unknown Fields' take also troubles the definition of fashion film. It does not easily sit as part of the documentary genre either, but it still participates in these questions about the intersections of the aesthetics of movement, fashion and contemporary debates in media and cinema studies. In this vein, I argue that fashion film is not merely limited to contexts of brand-driven post-cinematic aesthetics of marketing, but is part of the wider context of contemporary aesthetics. The aesthetic choices also feature the infrastructure of the technological culture of planetary production that establishes movement across borders while at the same time relying on specific locations and labouring bodies of colour.

My discussion in this text weaves together temporal aspects of garments, their aesthetic features and cinematic movement, and the conceptual themes of surfaces and folds that Giuliana Bruno, Anneke Smelik and others have brought into the transdisciplinary zone of discussions of cinema and fashion. As Bruno (2017) argues, this involves a shift from image to surface; in other words, a refocusing of the question of materiality where the 'theoretical interweaving of materials emphasizes the actual *fabrics* of the visual: the surface condition, the textural manifestation, and the support of a work as well as the way in which it is sited, whether on the canvas, the wall, or the screen'. One can observe the obvious relevance to discussions of fashion and cinema as arts of movement, but also, as I will continue, a further need to understand other scales of surfaces. In many ways, this chapter responds to Bruno's call to 'propose a different "model" for the theorization of fashion, one that is able to account for the way fashion works as a fabric of the visual in a larger field of spatiovisual fabrications' (Bruno 2014: 40), and to investigate how this approach to aesthetics and temporalities of materiality can also inform some ways of looking at moving images of fashion that are installed and shown in contexts of contemporary art and speculative design, and hence produce a different stage for the visibility of fashion film too. In other words, can fashion film also relate to the interesting challenge taken up in contemporary film and media studies – a shift that has focused on 'circulation, infrastructure, algorithms, labor cultures, technologies, and itineraries', as Elena Gorfinkel (2018: 123) has summarized it?

In *Unravelled* and the other examples of fashion film discussed in this chapter, the moving image becomes a point of articulation – and in some ways, a point of reproduction – of the logistical imaginary (Rossiter 2016) that includes aesthetic articulations of work and supply. Hence, what follows is a discussion concerning the concept of the surface and folds, followed up by a focus on planetary movement as the mobile site of image/fashion, fabrics of fashion, and a sort of alternative archaeology of fashion film that the installation produces: archaeology not as a historical map of forgotten paths, but as the conditions of existence of fashion items and their travels, where the cinematic movement produced by fashion films is itself conditioned by a planetary-scale movement across a surface that is one of logistics and distribution, of things being worn and things that wear out.

An alternative fashion film

Contemporary scholarly discussions about fashion film have highlighted its status not only as a recent genre of digital aesthetics and video, but one that has also longer roots in early cinema.[5] In the context of the 'Archaeology of Fashion Film' project (2017–19), we extended discussions from contemporary fashion films to early examples of fashion found on newsreels and in promotional contexts that can be discovered in little fragments of examples in multiple archives in Europe and the United States. Conceptually, such archival work benefits from the already existing rich vocabularies that have mapped fashion film not only as a genre, but as a thematic and aesthetic concern that can be seen as central to a cultural history of movement, spectacle, mechanization and mediatization of the (most often female) body.

Caroline Evans (2013) has articulated how bodily movement becomes controlled, defined and refined in relation to the cinematic image: the fashion mannequin as a cultural-historical figure relates to the media-technological contexts where techniques of close attention, new urban spaces and the spectacularization of the female body become enmeshed. The mechanical double of the mannequin and the human body in moving images becomes a key theme in the role cinema plays as production of movement. It expands questions of fashion to those of practices of moving as part of behavioural forms of managing the body where particular examples of movement – from Georges Méliès's 'amusing commercials for *Mystère* corsets' (Uhlirova 2013: 140) to the various instantiations since the 1890s of 'Loïe Fuller's notorious Serpentine dances' (Baronian 2017) to couturier Paul Poiret's promotional films and many examples of anonymous and sometimes-named models on-screen, from static to revolving, from close-ups to staged scenes – show the new fashion of fashion showing itself. Fashion is made visible in all the fundamental ways that pertain to the cinematic form of visualizing movement – or as Marketa Uhlirova (2013: 139) puts it, 'at a fundamental level, movement and change are the touchstones of the everyday "performance" of fashion'.[6] Furthermore, the question of cinematic contexts of fashion is where sight becomes a question of site and place, movement a question of transport and logistics: the particular situations, architectures and, as I will argue later, infrastructures that are essential for an expanded sense of what movement involves. Hence, the site itself is where scenes both domestic and public are brought into play, while the work of movement becomes part of it too (Bruno 2002: 15–16).

Sites are also part of the settings where cinematic fashion starts to exhibit bodies and where the work of bodies takes place. Cultural techniques of the body were managed both on- and off-screen, from stage to factory. While cinema was one central form of producing bodily behaviour and gestures (Väliaho 2010), it linked closely to the trope of factories too. In this context, 'factories of elegance' were not meant as metaphorical units of description, as Evans (2013: 144) also demonstrates. In the words of an anonymous couturier in 1923: 'My trade? It's in America that I learnt it best, from the big automobile manufacturer Ford. I apply industrial methods' (cited in Evans 2013: 145). As Evans goes on to show, the industrialization and mechanics of production, showcasing, exhibition and also embodiment in the female figure were all bundled up in what this particular early film archaeology of fashion included. Hence, there existed already a rich understanding of the mechanics of the body and the industrial production of fashion that was articulated through cinema. As for fashion film as part of this early modernity, one could observe the film as a crystallized expression of its industrial base.

There are various examples of both early and more recent fashion films that take the viewer 'behind the scenes', demonstrating the emergence of fashion as a visual spectacle and offering the process as the main feature of audiovisual pleasure. In a way, Alexander McQueen's *The Bridegroom Stripped Bare: Transformer* (SHOWstudio, 2002) would perhaps qualify as one example of the stylized – and exaggerated – performance of how a fashion dress comes about as a mix of creative bursts. However, in earlier films, this glance at fashion as a process could include examples that showcased the ateliers and backrooms, the manufacturing and the process of fashion: *From Wool to Wearer: The Romance of Pesco Underwear* (BFI, 1913), *Calais: De sleutel van Frankrijk* (Pathé Revue 26, 1924) and *Fabrication des gants* (Pathé Revue, 1924) count as some of the early examples. They articulate a cinematic interest in key drivers of modernity: machines, chemistry and labour, condensed in the spatial scenes and site of the factory.[7] Of course, in the core cinematic trope of 'leaving the factory', many short examples of workers exiting through the cotton mill gates in Lancashire and Blackburn and other places would count as curious examples of sites of work with fabrics and industrial modernity.[8]

If we extend the theme of 'behind the scenes' to contemporary factories and the industrial base in globalization, it is most often represented in documentaries that feature the exploitative conditions of the fast-fashion industry (cf. Hoskins 2014). Over the past ten to fifteen years, in films such as *China Blue* (Micha Peled, 2005) and *No Sweat* (Amie Williams, 2006), the exploitation of labour, both in the United States and in outsourced locations such as China, is brought into focus and extends as part of the work in which many non-governmental organizations have engaged. Similarly, Meghna Guptna's short documentary *Unravel* (2012), about the return to India and recycling of fashion items as reused fabrics, becomes part of the extended discussion in ways that also resonate with *Unravelled* in more than name only. Furthermore, as Nick Rees-Roberts (2017) points out, *Unravel* and works such as Jia Zhangke's *Useless* (2007) are necessary counterparts to the usual aesthetics of fashion film, as they extend the discussion of both material and symbolic forms of desire to the planetary contexts of fashion industries. More recently, Rahul Jain's *Machines* (2016) moves into a related terrain with carefully composed long, slow shots, a powerful minimalist machine soundscape, and a mix of factory interiors where machines, textile and labour mix into multiple levels of cinematic patterns. The more stylized and cinematographically skilful examples of work clearly continue along the key intersections between sites (factories), labour, movement and fashion as it becomes expressed as material and as desire.

Something similar is visible in *Unravelled* too, and in general is present in how Unknown Fields' work can be characterized as a practice of topoi, mapping places and spaces as defined by their routes and connections, as Casey Boyle (2018: 152–5) argues. Images shift from factories to landscapes, following a route that extends the view of the production process to the environmental elements. Surfaces of fashion – as images and as environments of images – are related to multiple ecologies, indicated in my quotation from Unknown Fields' own words on page 21: rivers to screens, cloth to skin, labour to logistics and the visual surface from cinema to digital. The map is the surface is the screen is the territory wrapped as a fabric that moves along the planetary conveyor belt (Young 2015b) from factories to cities. The expedition and the images in the installation are maps of the planetary surface movement that functions as a moving image in its own right: they articulate natural resources as part of labour, labour as part of space, and the geographically widespread connections that fashion establishes as a material practice. Imagery of fashion is displaced from its usual contexts as featured in most contemporary cinematic examples: short music-video-style clips, often for online distribution, and featuring increasingly famous directors such as Spike Jonze, David Lynch and Wes Anderson. However, in this lineage of the cinematic short form, *Unravelled*'s context of an art gallery is not completely out of place, considering how, in Bruno's (2014: 62) words, 'contemporary art is mining, reshaping, and circulating an archive of cinematic experiences'.

Even if Unknown Fields' work is contextualized primarily as part of speculative design and architecture, there is a relation to many forms of audiovisual expression in post-cinematic culture. Many of the past years' research expeditions have come out as short, stylized audiovisual showreels and clips, with *Unravelled* being the most finalized, gallery-friendly product that also most clearly sits in relation to earlier audiovisual traditions – namely, indeed, fashion film. In *Unravelled*, movement and change are presented as an audiovisual piece that also stems from design research expeditions on the move as part of the international transport of people and images.

In this sense, I want to pay attention both to the condensation of themes that pertain to the fashion film as a recurring cultural-historical aesthetic of

Figure 1.4 *Unravelled* (film still), *c.* 2017, directed by Unknown Fields in collaboration with Tushar Prakash.

folds, surfaces and movement, and to how movement relates to geographies, planetary surfaces and logistics and labour expressed in the audiovisual installation. The design research methodologies and speculative themes are ways to open up fashion film from outside the industry and place it in a broader critical set of contemporary debates on cinema, media theory and aesthetics.[9] As part of this work, I will follow the direction Bruno set out in her work in both *Atlas of Emotion* and *Surface*: contemporary screens are mapped out in relation to manifold surfaces, the materiality of screens as textures, and textures as they are stretched across moving images that are more haptic than representational. This pertains to the work of the 'archaeology of migrant media' (Bruno 2014: 6–7): tracking the surface movement across architecture, textile/cloth and the (post-)cinematic. But I am as keen to move it in a new direction that concerns multiple other scales of surfaces, including planetary infrastructures, and as such the archaeology of migrant media starts also to include a necessary link to an archaeology of migrant labour – and the way critical studies in fashion have pointed out this dimension of the industry.[10]

The surface fold

Unravelled works by way of multiple folds: the long golden drape of a sari-like cloth occupies space in ways that emphasize its folds as part of that space even while they move across it. But across scenes, folds multiply: stacks and packs of cloth, fabrics pressed and lined up, machines operating upon material, and even the screens as a fourfold operation across which the multiple simultaneous scenes work in combination. And across the four screens of the installation, multiple rhythms and materials appear simultaneously: there are threads, cotton plants, machines and movement; there are fabrics, cloths, fashion walks and sewing; there are stacks of cloth, coloured folds, hands at work and the soundscape that accompanies the image – the rhythmic click and clack of machines organizing the tempo of work. The soundscape of the repetitious click is what binds factories, film projectors and typewriters together as forms of mechanization and modernization[11]: the factory itself is part of the same cinematic lineage where images, like clothes, are mechanically produced into existence.

To understand how many of these visual themes are connected, Bruno's conceptualization of the fold is particularly useful. The fold marks a connection between cinema, fabric and architecture, as well as embodied movement (see also Smelik 2015). In Bruno's (2014: 22) words, 'the fold is a piece of geopsychic matter that is an actual element of fashion' where the folded surfaces of fashion garments embody an active sense of agency that is, however, not disconnected from the particular skins they are in touch with: as they touch bodies, they also touch and mark topographies as maps of movement that are expressed in how they are worn, or even inhabited.

But it is important to extend to geographical movements, planetary scales, fashion in architectures, and images that move to show the infrastructures of their own existence. The surfaces cover more than individual bodies or even architectures. The focus is extended into layers of surfaces, around which Unknown Fields' video installation also frames the issue. Bruno's interest in visuality as a haptic surface condition involves a conceptual and aesthetic thickness, but one that also layers histories: 'One can say that a visual text can even *wear* its own history, inscribed as an imprint onto its textural surface' (Bruno 2017). It is in this context of multiple modalities of thickness expressed as folds (of images, garments, surfaces, histories) that this chapter maps postcolonial aspects of the planetary nature of surface and the aesthetics of fashion film. This thickness helps to rescale the focus that is inspiringly present in Bruno's work from architectures to infrastructures, from the urban to its conditions of existence, from cinematic screens to the large-scale movement that is established in what is framed in *Unravelled*: the global logistics that is the extended factory of cinematic movement of fashion garments.

In other words, there is more to the surface than meets the skin. One part of this extension is the architectural. As Bruno (2014: 30) reads through Wong Kar-Wai's *In the Mood for Love* (2000), 'fashion is an architecture' that constructs folded depths on surfaces and connects material and psychic textures as atmospheres, moods and affects. These surfaces are radically temporal in the sense that while being inhabited – as fashion, as architecture, as houses of sorts – they are also worn out, as Bruno elegantly puts it, referring to the inscriptions left by the passage of time. Hence, importantly, surface as it is expressed in the triangle of fashion, film and architecture goes 'much beyond costume and becomes an altogether different object for the circulation of meaning' (ibid.: 39). For Bruno, this particular form of desire is evident in Wong Kar-Wai's film, but one must specify that it is not about singular objects or cinematic representations, but about the circuits of materials across multiple scales of texture and process.

Surfaces connect disciplines. The conceptual territory of the surface becomes part of a multidisciplinary methodology of investigating multiple scales of embodiment. While the terminology of the surface might itself become too broad or vague, as it can so suitably speak to multiple contexts, it definitely is useful to address it as interstitial, as Tim Ingold (2017: 104) has proposed. Surfaces fold conceptual topographies across architecture, screen studies, design and fashion, and speak to how the visual extends to the haptic. The surface connects disciplines across a range of methods and also necessitates the development of methods. The surface 'method' – whether pertaining to fashion studies (Bruno 2014: 39) or broadly to this multidisciplinary audiovisual encounter with scales of materiality – would not, then, be a question of superficiality and lack of depth in the sense intended by early media and film critics (see Kracauer 1987), but would relate to the interstitial spaces that are formative of surfaces (again echoing Ingold), and to cartography in the manner that Gilles Deleuze developed reading Foucault, and which Rosi Braidotti (2006: 11) then developed further: cartographies map historical conditions; they pay attention to localities as parts of nexuses of powers, where forms of thinking through topologies meet up with forms of governance, and where the space is always folded. This also links to understanding the planetary not as a homogeneous space of exchange, but as a localizable space of alterity, in Gayatri Spivak's (2012) terms.

The cartographical surface tells both of planetary travels and territories and also of the folds that tie in psychic, affective and cinematic realities in other ways. Through Bruno, the fold is mobilized from earlier discussions by Deleuze into the contemporary context of aesthetics. Insides and outsides are always surfaced through the fold, and surface is already a fold of insides and outsides.[12] Hence, to move along the surface entails asking: how many routes are there, and through which milieus, contexts and materials do the routes proceed? How many interstitial surfaces enter into our accounts that address contemporary materialities, contemporary forms of knowing and governing surfaces, and ways of making surface media, or media to enable us to map and see a surface? How do maps, screens, clothes, skin and labour start to speak to each other, and even to the other sites that pertain to the sense of the planetary as it is discussed in recent debates about the Anthropocene, which was also the theme of the exhibition mentioned earlier? From the mapping of the planetary surface to the textile surfaces upon our skin, the fold connects: it is a multi-scalar relation that speaks to this topological theme where things at our fingertips are on a planetary scale.

Cartographies, architectures and films share much in common. One theme is about space and its connections. The particular attention to sites of film – and in our case, the fashion film as it moves from the showcasing of fashion as a product to the production industry, logistics and planetary movement as a commodity – links to how film has in many ways already functioned as cartography (Conley 2006). One form of the archaeology of moving images would then recognize this intimate bond with how sites of travel, tourism, transport industries and also city life are singularly expressed in the cinematic aesthetic, even amounting to a form of psychogeography (Bruno 2002: 276–7). But the other, complementary way to understand the work of mapping and surfaces is to be aware of how maps were already also cloths: they functioned as an important infrastructure of trade routes as well as colonialism, facilitating movements of cotton and cloths, fashion and fabrics, and for a long time they were also materially and metaphorically textured as cloth. As Veronica Della Dora (2015) shows, geography as inscription also includes the materials on which inscription takes place. Rather than focusing on the cinematic form of inscription (which itself is central to how maps, textures and materials are entangled), she points out the haptic sense that was part of the understanding of maps for a long time: in both 'Hellenistic and Judaeo-Christian traditions … the earth and the heavens repeatedly feature as garments, thus marking the history of western (geographical) thought for the centuries to come' (ibid.: 226). These could also mark the multiple scales of movement – such as in early Christian depictions that use textual forms of folding and clothing as a way to move from a 'macro-scale of the cosmographic mantle' to how it 'is mapped on the micro-scale of the surface of the suffering body of Christ' (ibid.: 229). Bodies, maps and cloth are tightly pleated.

While the term 'mantle' has often featured as a synonym for surfaces, Della Dora points out the persistence of cloth for understanding not only geography as an inscription of place but also its material dynamics. The medieval *mappa mundi* serves here as direct reference to the 'cloth of the world' (ibid.: 233), while multiple kinds of fold are continuously part of this materiality of maps. For example, 'the "unveiling" motive recurs on later frontispieces of prestigious geographical works, including early eighteenth-century editions of Strabo's Geography' (ibid.: 237). While the complexity of modern geographical forms of surface itself becomes a function of its media-technological and geopolitical mapping, the generic theme of this fabric of the map sits interestingly as part of the cinematic starting point Bruno offers. Here, the fold and its materiality are ways to express scales, similarly to the way textiles as woven structures are both minute and radically expandable. Furthermore, and importantly for the argument developed in this chapter, 'mantles, curtains, surfaces and veils all operate as points of contacts and as points of tension between the visible and the invisible, between the micro-scale of the Self and the macro-scale of the world' (ibid.: 245–6). But instead of the self and the world, I am interested in how figures of fashion and textile are points of contact across space and scale, and extend to include the labouring bodies of fashion as a global industry. Continuing the discussion of how space and the moving image are points of such tension, the next section returns the focus to *Unravelled*, including a discussion of how the film is involved in a fold of multiple materials that also include patterns of images of fashion as work, and sites of moving images such as the factory, a recurring cinematic trope.

Garment planes

In a similar way to *Unravelled*, the documentary *Machines*, briefly mentioned above, features audiovisual contrasts between folds of stacked fabrics, factory spaces and their repetitious soundscape. It also features multiple cinematic series that structure the film: scenes of machines and movement; scenes of textiles and folds (including colour-mixture processes); scenes of exhaustion. The theme of the three pillars of modernity (machines, chemistry and labour), found even in earlier examples of fashion film, features in this factory too. However, *Machines* is shot in Sachin, in Gujarat, India, and, as such, it too includes an implicit historical fold: the region's role in early fifteenth- and sixteenth-century global patterns of merchants and trade in cotton, and contemporary forms of the global production of textiles, fabrics and fashion. Indian subcontinent cloth also travelled through the networks of Gujarati merchants before the colonial Portuguese and other trading companies arrived: 'The community of merchants that stretched from Egypt and West Asia to Melaka and Java created a shared market, a trading diaspora whose reach is evidenced by the many Indian trade cloths discovered in Southeast Asia that are analogous to those found in Egypt' (Guy 2013: 15). The transport of dyed cotton cloth and the desired patterned textiles established what historians have dubbed 'threads that bind' (Machado 2009: 161) and 'webs of materiality' (ibid.: 179) – such apt metaphors considering the ties between particular locations, their relation to planetary trade routes, and the haptic theme of materials that become part of interior décor, clothing and the global travels of motifs and patterns in general. 'Woven in myriad styles and patterns, these cloths were central to material exchange in the ocean and were quite literally the threads that bound consumers to one another on the one hand, and producers, consumers, and merchants on the other' (ibid.). Maps and cloth become two sides of this haptic materiality of movement and infrastructure: the merchandise and its routes mapped surfaces that were essential for the linking and governance of territories. As Peter Sloterdijk (2013: 100) sums up, 'imperialism is applied planimetry: the art of reproducing orbs as surfaces and worlds as charts'. The planetary cloth of the map allowed the movement of fabrics, cloths and the *longue durée* of fashion.[13]

This scalar-hopping is also what implicitly defines the otherwise slowly unfolding narrative movement of *Unravelled*. Movement spurs movement: the cotton mobilizes factories, mobilizes work, mobilizes threads and fabrics, mobilizes dye, mobilizes clothes, mobilizes the movement of fashion that is recursively about movement and embodiment. The images of transport boxes alongside the maritime references bundle up the film's other series of images as part of what Liam Young (2015b) of Unknown Fields has called the planetary-scale conveyor belt. The physical locations listed earlier are tightly glued together by the cinematically established movement across spaces on screen that both map discrete parts of production and demonstrate their links – the moving image works here as an articulation of what logistics does in terms of space: maps establish temporary, operational continuities.

Unravelled forms a link to Unknown Fields' earlier work too. One can relate it to the broader infrastructural turn in critical design as well as in media and film studies.[14] Unknown Fields' speculative architectural angle addresses non-urban infrastructures – the 'dislocated city' – including methodologies that perform the role of what is also being investigated: data-intensive ways of understanding planetary movements of goods and information, the governing role of logistics, and the conditions of existence of technological culture. The various earlier expeditions have engaged with energy (oil and north Alaska, and the role of the Global Positioning System (GPS) as an enabling technology; lithium-mining in the Atacama Desert in Chile and the Bolivian Salt Lakes; and the geographical sites of emerging technological futures such as electric batteries); transport (for example, the winter 2013 project 'A World Adrift', which moved on the surfaces of maritime routes and the architectural sets that define this space and its vectors of movement); and luxury cultures (economies and labour of gem-mining in Madagascar[15]; and the 'Rare Earthenware' project, which used mineral-processing waste to create a mock-luxury object for the V&A Museum in London). While the rhetoric of their expeditions as travels to 'alien landscapes, industrial ecologies', 'dislocated landscapes [that] are embedded in global systems that form a vast network of elusive tendrils, twisting threadlike over everything around us, crisscrossing the planet, connecting the mundane to the extraordinary'[16] can easily sound like an overly romanticizing trope of anthropological fieldwork,[17] it points to the extended scales which define much of what takes us outside the usual visible surfaces of film, design and media culture. Of course, all of the above about the planetary conveyor belt and optimized logistics fits closely with the industrial backbone of fast fashion, where issues of the exploitation of labour (including migrant labour) and environmental hazards are not some hidden secret to be revealed, but are already well known and often articulated in both research and public debates (Hoskins 2014). In other words, 'making something visible' in terms of the infrastructure of such planetary industries is not automatically a liberating act in itself, but has to be understood as part of a wider circulation of images about labour and infrastructure.

Unravelled is Unknown Fields' first piece to be produced as a gallery-style moving-image installation. As argued above, it also works as an example of an alternative fashion film that extends the reach of both critical design methodologies and fashion discourse into the contemporary art context as well as discussions of the Anthropocene. It articulates a very different set of moving images from the blockbuster scale of *Anthropocenema* (Kara 2016), but it does work with themes of scale and planetary movement in its own subtle way. Located in specific sites of work and experience that include the long historical traditions of manual expertise with textiles and dyes, and establishing how they are constantly glued, articulated, linked and composed as part of a movement of fabrics and images on a planetary level, it expresses part of the moving ecology of fashion. Including the labour of people of colour might itself be one extended way in which the visibility of bodies in fashion film becomes rearticulated, while it also participates in the cinema of logistics of planetary-scale factories of image/fashion.[18] To refer to the planetary as a contrast with globalization has – as argued by Spivak (2012) – become a way to emphasize the localizable situations, bodies, labour and even friction that set the seeming smoothness of the global in action, and support it. Hence, it is necessary to take literally Spivak's framework for the planetary: how to reimagine the planet in its infrastructures, sites and situations that relate to the planetary conveyor belt. What sorts of image sufficiently address the multiple scales that overlap, from labour and localities to the mechanisms of transport that logistically ensure the seeming smoothness of movement of materials such as fashion garments and apparel?[19]

To conclude, *Unravelled* does expand the scope of moving images of fashion, while at the same time benefits from being able to move along the very vector of the logistical imaginary that defines it: the circuit of fabrics is the circuit of images of fashion; the showcasing of fashion styles as the blunt function of fashion film since early cinema (D'Aloia, Baronian and Pedroni 2017: 7) is the showcasing of images of what produces fashion styles; the planetary circulation of genres, types and gestures in images is parallel to what fashion does anyway. This context provides a further site of articulation of the planetary as one reference point for moving images of bodies in motion (ibid.: 5) and garments in motion. And it is this multiple scale of the ecology of movements that is discernible in the way Bruno's take on surfaces can be expanded to touch contemporary discussions about the planetary: surfaces of maps as vectors of possible movement, surfaces of images as articulations of bodies of fashion spectacles and bodies producing those spectacles, and surfaces of architecture not merely as clothing, as Le Corbusier once suggested (Bruno 2014: 55), but also as infrastructure. Movement, textile and cloth fold into the metaphors of interwoven global surfaces, the maritime routes, the maps that enabled the systematic coordination of ships, materials and products. These are expressed in a moving image of a surface distributed across the digital landscape that is, of course, not merely about a screen of representation, but technological forms of enabling movement: the logistical backstage. Hence, the design work and mobile studio of Unknown Fields offers examples of an audiovisual product that uses the aesthetic connotations that pertain to a genre (fashion film) but moves it into its own investigation of the planetary surface of the Anthropocene, which marks the displacement of that singular human perspective in favour of other ways of mapping movement across surfaces: movement moving, and the logistical imaginaries of bodies in the movement of work.

Notes

1 The research for this chapter was funded by the Arts and Humanities Research Council ('Archaeology of Fashion Film', 2017–19, AH/P004598/1 project).

2 'After the End of the World' (25 October 2017–29 April 2018, Centre de Cultura Contemporània de Barcelona). http://www.cccb.org/en/exhibitions/file/after-the-end-of-the-world/224747 (accessed 23 April 2018).

3 See also Casey Boyle's (2018: 155) characterization of Unknown Fields' practice as a topoi study that is interested in infrastructures of connection vis-à-vis architectures of containment.

4 'Winter 2017_Unravelled', Unknown Fields website, http://www.unknownfieldsdivision.com/winter2017india+bangladesh-unravelled.html (accessed 8 May 2018).

5 As Marketa Uhlirova (2013: 139) puts it, there is a distinct benefit 'to look[ing] deeper into the past and to trac[ing] many of the fashion film's key characteristics back to the "pre-digital age"', which includes a conceptual focus on the moving material of cloths and garments in resonance with the movement of images.

6 Consider also Esther Leslie's (2013: 34) phrasing: 'Film is a medium of movement and so provides a perfect vehicle for showcasing cloth's mobile propensities. In turn, cinema needs the drama of movement.'

7 Later examples such as *Making Fashion* (Humphrey Jennings, 1938) also open up to the process of making fashion, but are less about the industrial setting and more about the romanticized creative process of design.

8 For example, *Workers Leaving Haslam's Ltd, Colne* (1900), *Workpeople Leaving Ordnance Mill, Blackburn* (1900) and *Employees Leaving Williamson's Factory, Lancaster* (1901), all found at the BFI. Cotton factories in the north of England were a steady theme in many newsreels and documentary-type products for decades.

9 One interesting theme also concerns fashion film as post-cinema, but that is unfortunately outside the main scope of this chapter and has to be left to be discussed in another context.

10 See for example the SOMO 2016 report 'Migrant Labour in the Textile and Garment Industry'. https://www.somo.nl/migrant-labour-in-the-textile-and-garment-industry/ (accessed 16 July 2018). See also Fletcher and Tham (2016) and Ross (1997). In addition, cf. the 'Global Circuits of Fashion and Beauty' event that took place at New York University in February 2015. http://apa.nyu.edu/event/global-circuits-of-fashion-and-beauty/ (accessed 16 July 2018).

11 This is a reference to Vilém Flusser's essay 'Why Do Typewriters Go "Click"?' (1999: 62–5). Laura Mulvey (2006: 50) also refers to the click of the high-heeled shoe of the flapper as part of this cinematic modernity. Mechanization and the female body are the key themes in Evans (2013), which also develops Mulvey's point.

12 To quote Deleuze (1988: 119): 'Between them there is a topological relation: the relation to oneself is homologous to the relation with the outside and the two are in contact, through the intermediary of the strata which are relatively external environments (and therefore relatively internal).'

13 Hence, from the maritime routes of early trading companies such as Carreira da Índia and the East India Company, among others, to the more recent forms of what Keller Easterling (2014) calls extrastatecraft, the territorial and maritime control of surfaces defines this other scale of surface-space.

14 See, for example, Parks and Starosielski (2015) and Rossiter (2016). Peters (2015: 33) defines infrastructuralism in the following words: 'Its fascination is for the basic, the boring, the mundane, and all the mischievous work done behind the scenes. It is a doctrine of environments and small differences, of strait gates and the needle's eye, of things not understood that stand under our worlds.'

15 One key material narrative in that expedition is the metabolic transformation of food into work into luxury artefacts: 'After manufacture, the gemstone has been set into a gold tooth, ready for that million-dollar smile and the outrageous lyric. From kilojoules, to carats.' Project: http://www.unknownfieldsdivision.com/summer2013madagascar-journeytotreasureisland.html (accessed 16 July 2018).

16 Unknown Fields' 'Mission'. http://www.unknownfieldsdivision.com/mission.html (accessed 16 July 2018).

17 As Shannon Mattern (2016) points out, this is not entirely unproblematic when one realizes that the rhetoric reproduces elements from the genre of the colonial expedition – mapping uncharted territories – and recycles masculine tropes of rite of passage in travels through rough terrains.

18 But, of course, it might always also risk making the labour of coloured bodies into a spectacle, or part of the media and artistic critique that remains on the level of 'making visible'.

19 These questions of planetary images resonate with some work in the context of sustainability and fashion (see Fletcher and Tham 2014).

List of Figures

Chapter 2

Surface-making in Nuclear Decommissioning: A Narrative of Sludge, Plutonium and Their Whereabouts

Petra Tjitske Kalshoven

A visit to a legacy pond: just below the surface

In July 2018, almost a year into my fieldwork at and around Sellafield,[1] a visit was arranged for me that would finally get me into the SEP (separation) area, behind additional fences and turnstiles, on this nuclear site in north-west England. A conversation with the head of legacy ponds had led to an invitation to the 'less dosey' of the two legacy ponds[2] on-site, the Pile Fuel Storage Pond (PFSP). Following my escort into the separately fenced-off area gave me the rather exciting impression of penetrating another surface of the heavily secured site, where, as I will argue, surface-making is key to its mission of containing radioactive wastes.

Before we could enter the pond building, protective clothing was prescribed, which I had not been asked to wear on the site before. My escort, a female chemist who explained to me that she led the operations support team for the pond, accompanied me to the changing rooms. Only 'basics' were needed for the visit, including a dark-blue shirt and trousers, safety shoes and bright-orange socks (considered an unattractive and unmistakable colour, so they would not get nicked). An engineer joined us, the operations manager. While still outside, he went through a safety briefing with me, ensuring I did not have any open wounds or scratches on my skin, which would have made me vulnerable to radionuclides. They escorted me into the building, and then out to the large open-air pond. The walkways were narrow, requiring a constant stepping upwards or sideways to avoid obstacles in the way. This was quite a change from the smooth, brightly painted surfaces I had come to know in other nuclear buildings, surfaces that are easily wiped off if contamination should occur. Here, it was the water acting as a barrier, keeping radionuclides at bay until the waste that for decades had been left lingering in the pond would be retrieved and made safe.

We walked clockwise around the open-air pond. There was a slight drizzle. I wondered whether this made the radionuclides stir under the protective surface of the pool. I peered into the water but did not see much. The ops manager said that is because of the reflection of the sky. He pointed out a few things, such as the skip tipper of which the lid had become stuck – I could barely make out some whitish shape. It would have to come out, I was told. Everything would have to come out. Going back in, we found the maintenance team preparing for this job, working on what looked to me like a myriad of tubes; using different dyes, they had located a leak in a manifold. On the outside of the basin, down on the 'quay' as I thought of it – it felt as if we were on a ship – lay large white plastic bags filled with metal debris. The chemist explained to me that these bags had been manufactured abroad, made to measure, which was very helpful, for example, to wrap the bay doors that had been taken out – the ops manager pointed out where these had sat before. The guys, said the chemist, get less exposed to radiation if these bags can be quickly zipped up. Zipping up was done with long poles to put distance between the operators and the 'inventory'. It was her team's responsibility as ops support to organize and procure such bags.

On one side of the pond, a man was getting into a suit that looked as if it was made out of paper: this is a rigger, the ops manager told me, do you know what he does? Riggers hoist equipment, he explained, and men like G have spent years figuring out ways of getting jobs done. In this case, G was aiming to retrieve trolleys from the pond. He was going to use the yellow crane or skip handler sitting at the right end of the pond, and a lot of blue tape and blue rope, which, I was told, would be put in place through a plastic tube underneath a trolley-to-be-retrieved, and he would use any pipes he could make useful, while an ROV (remotely operated vehicle) floated around in the pond to guide him. We chatted for a while. What struck me most was how G wrapped the blue tape around his wrists to make the transition between his suit and his gloves seamless – it looked so very improvised and yet was effective, I gathered, working as a seal to prevent the spread of contamination through the seam.

The visit over, I was escorted out of the SEP area and made my way back to one of the Sellafield site gates that I was able to negotiate by myself, scanning my pass and moving through turnstiles, leaving the security shell – or so I thought. At the very unassuming Sellafield railway station a short walk away, where the river Ehen runs parallel to the Irish Sea, the presence of the nuclear site had spilt over: two Civil Nuclear Constabulary men in armoured suits patrolled our platform. A young man standing next to me, who hoped the diesel train would not come in too late for his connection in Carlisle, struck up a conversation with one of them, a personable yet very intimidating-looking tall man cradling an assault rifle. He felt rather hot, he told us, wearing three stone of gear – and that was excluding his weapon. It was the plates in the body armour that were so heavy, he explained, plus the ammo.

Sellafield and surface-making

Recent scholarship on surface emphasizes the layering that underlies and creates surface, implying processes of sedimentation over time. Because of the temporal peculiarities of radioactivity, which, in its different material manifestations, extends backwards and forwards through timescales that are beyond imagining, 'the nuclear' is eminently subject to processes of sedimentation yet stubbornly resists containment in any specific layer. Humans engage in processes of multiple layering with a view to containing radiation, making it safe, secure and acceptable, creating surfaces of materials, technologies and discourses that seek to keep it in check.

In both the discourse and practice of nuclear waste storage, 'surface' is a contentious place and concept because of its association with porosity and thus the risk of a lack of containment. Nuclear waste kept 'on the surface' must be made safe through packaging in dedicated materials that create additional layers of surfaces around radioactive substances. The Sellafield nuclear site in West Cumbria is a place where surfaces are constantly in the making. In its treatment of nuclear waste that needs packaging and repackaging, in its protection of humans and the environment, in its core business of containment, Sellafield Ltd's (SL) main method and occupation consists in surfacing and resurfacing. Surfaces are added to the body: barrier cream on skin; leaded or zinc bromide glass that protects eyes looking at radioactive samples; protective suits to shield against radiation; armoured suits for security. Surfaces are created by packaging and repackaging waste: through vitrification to contain high-level waste (HLW), through encapsulation of low-level wastes (LLW) in concrete which is subsequently secured in boxes, drums, containers, flasks – all practices meant to provide additional surfaces to waste products, thus creating wasteforms.[3] Research into methods and materials to perform these tasks in the best possible way is a lively matter, partly supported and funded by the Nuclear Decommissioning Authority (NDA) in collaboration with universities and research institutes, the National Nuclear Laboratory among them.[4] The most sophisticated, innovative high-end research occurs in direct conjunction with the very hands-on (and yet profoundly hands-off) making-do that I witnessed at the legacy pond.

All the surface-making that occurs at Sellafield is meant as a temporary measure to prepare for a long-term, even permanent solution to dispose of nuclear waste in a so-called geological disposal facility (GDF) that does not yet exist in the United

Kingdom.[5] In a piece discussing nuclear waste storage at Sellafield and the policymaking aimed at permanent disposal, geographer Nicky Gregson (2012) has drawn attention to the vitality of nuclear stuff, to its 'thing power', which she feels is insufficiently acknowledged in technoscientific discourse. In this chapter, I will revisit Sellafield and the vibrant matter it houses in the context of the UK government's recently renewed quest for a GDF, to be located deep under the earth's surface. Focusing on the surface-making that I argue is key to nuclear decommissioning, and drawing on my ongoing ethnography at and around the Sellafield nuclear facility, I touch upon technologies of containing (packaging, storing, burying) and the materials involved in these technologies (radionuclides, as well as materials for their containment: concrete, glass, engineered barriers), with special attention to two specific materials (plutonium and sludge) that resist straightforward categorization and thereby disturb conceptualizations of a neatly bounded surface. Drawing on ethnographic engagement with the nuclear industry in its West Cumbrian context, I argue that unruly materials may be allies rather than embarrassments in their potential to enliven public debate on nuclear decommissioning as a practice fraught with unknowns and uncertainties.

Context: a Sellafield storage pond

The West Cumbrian Borough of Copeland is home to the Sellafield nuclear site, which has a rich and controversial history. It began its nuclear involvement as a site for the production of plutonium for a British atomic bomb. It produced nuclear power from 1956 (when the first commercially operated power plant, Calder Hall, was opened) until 2003, and it has been actively reprocessing spent fuel for both domestic and foreign customers. This latter activity is set to end in 2021, and from then on Sellafield will be fully focused on decommissioning. The company in charge, SL, has since 2005 been a subsidiary of the NDA, which is a government body.[6] The NDA currently spends almost £2 billion per year on Sellafield's decommissioning, which is expected to take more than a hundred years – taking down buildings, decontaminating structures and 'remediating' the environment.

Special attention is paid to a number of so-called legacy facilities in which nuclear materials have been placed over the years, with little care or foresight spent on their future retrieval. In some cases, very little is known about the sediments of waste that have built up over time, making retrieval a hazardous and slow process. A number of legacy facilities have been labelled 'intolerable risks' and are considered priorities in SL's clean-up. Penetrating the protective surfaces that contain nuclear waste as a first step in retrieval operations, aided by robotics, is an uncertain and thus risky undertaking.[7]

Against this background, the PFSP is presented as a success story. Organized under the 'retrievals' value stream,[8] the pond offers the unity of scene (but not of time) required of an Aristotelian plot: a neatly delimited facility enlivened by an exciting narrative, a detective story of problem-solving, making-do and progress. After my visit to the pond, my host at the facility told me about the importance of trying to gather information about its contents: the more known about these, the more confidently and effectively the retrieval of materials can be achieved. He showed me a diagram of the pond on his computer, where each skip (all in different colours and codings) had a 'passport' attached to it, with information copied from handwritten notes from the 1950s; the diagram was updated daily as additional information about the skips' contents became available.

All in all, about 120 people were deployed on this pond. It was quite a varied group, the ops manager told me, ranging from young apprentices to people in their sixties. The variety of the work made it quite attractive. For instance, they would search for solutions on the internet regarding how to kill algae to improve pond visibility (increased activity as a result of decommissioning causes disturbance) and then call suppliers of algae killers. They would use low-light cameras and sonar, learning from experience and experimentation how to operate differently. So, is innovation about trying things out in new environments? I asked. It is about trialling, he said, like introducing a 'knock-out pot' (used in oil and gas and chemical industries) for centrifuging.

Progress reports on the pond, together with those on other legacy facilities, feature at academic and industry conferences, including 'TotalDecom' in Manchester (May 2019), where the ops manager presented. They are a recurring feature at the Risk and Hazard Reduction and Waste Management Working Group of the West Cumbria Sites Stakeholder Group (WCSSG).[9] In March 2018, the PFSP programme manager presented an update to the group. One decontamination effort that he elaborated on involved the creation of new surfaces by removing old ones through so-called pond wall shaving trials. He explained that they would shave off 3–9 millimetres of concrete to ascertain how deep the radioactive contamination sits and to gain an understanding of how concrete behaves. Having proved successful with a mock-up, the method is to be tried out on one of the bays after the water has been removed. The aim, we were told, is to have the entire pond dewatered by 2029, a time period that testifies to the time-consuming nature of nuclear decommissioning.

Sludge as an elusive presence

Heavily featured in the presentation for the WCSSG was sludge, the most radioactive component in the pond, consisting of algae and erosion products. The manager explained that the team worked to get the average concentration up from 6.5 per cent to 10 per cent, to try to minimize the number of drums required to store it – space is tight and costly at Sellafield. Drawing on operational experience as a learning process, he showed us an image of a worn impeller eroded by sludge, insisting it was vital to learn about and understand sludge characteristics. To complicate matters, in the sludge-retrieval area, blocking and clogging occurs. Sludge tends to hide piles of debris, he added, evoking a vivid image of the pond environment as a space of sedimentation. Lots of odd things are found in the pond, he told us, including wellies. Most debris is categorized as low-level waste (LLW) or intermediate-level waste (ILW). LLW is sent to the low-level waste repository (LLWR, see p. 43), while ILW is transferred downstream to treatment plants for packaging. Retrievals are assisted by ROVs and a remotely operated micro-digger, all meant to create distance between operators and radionuclides. The focus is on taking out a number of skips that were lowered into the pond in the past, so that room is made to work on removing sludge. Materials need to settle down, and this takes time.

During my visit to the pond, sludge behaviour was also commented upon as something quite puzzling. It was as if sludge behaved differently once it had been moved into the corral, my hosts said. They seemed to lose sludge in the process, finding much less of it present in the corral than they thought had already been moved there, not knowing where it might have gone. Figuring out sludge behaviour, the chemist said in exasperation, was like a 'black art'; it felt as if they were 'chasing ghosts'. At the next meeting of the WCSSG Risk and Hazard Reduction and Waste Management Working Group, a year later in March 2019, the same programme manager that had previously reported on the PFSP confirmed the misbehaviour of sludge. Many sludge exports have been put on hold, he told us, 'because we were just chasing sludge around the pond'. The team decided instead to concentrate their efforts on lowering the pond's water level, so that de-sludging could be pursued with more efficacy. Dewatering of the pond was also helpful for the continuing project of concrete-shaving, which, he explained, is done handheld above the water. 'The pond needs dewatering because shaving must be done on an exposed wall, otherwise the water will contaminate the wall again.'

Decommissioning as exploration below the surface

Accounts of cautious trialling in territories not explored before are quite typical in conversations with scientists and engineers involved in the decommissioning of Sellafield legacy facilities. Opening up what has been kept contained in facilities that are no longer considered fit for purpose implies risking a temporary porosity with a view to enhancing containment for future years to come. If you move into uncharted territory, discussion partners emphasized, you are bound to encounter the unknown and need to respond accordingly. Because of unexpected technical challenges, including strange behaviours of materials, targets and deadlines are difficult to set and achieve. Official promises to speed up the clean-up were seen as frustrating – there were just too many uncertainties. Such candid acknowledgement of the limits of what is possible stand in some tension with the consistently positive tone that is adopted in official communications, such as on the NDA's website, which generally presents upbeat highlights aimed at maintaining a smooth surface.[10] While decommissioning in practice is about breaking surface, its public face is about a sustained reassurance of competent containment, underpinned by regulatory scrutiny by the Office for Nuclear Regulation and the Environment Agency.

Storage versus disposal

While trials pinning down elusive sludge continue in situ and in simulations, a constant concern is ensuring the most effective waste routes – that is, knowing where the waste, once captured, can be sent. SL's LLW is transported to the LLWR near the village of Drigg (which is south of Sellafield, also in Copeland Borough), while ILW and HLW are stored on-site – as it is called, 'on the surface' – awaiting a disposal route.[11] Packaging and repackaging of stored nuclear materials, particularly of the higher-activity wastes (HAW),[12] is done as a temporary measure pending a more definitive solution. This solution – and we are told that it is generally internationally scientifically agreed that for now this is the most appropriate solution to make HAW secure *permanently* – is to bury it deep under the earth's surface through so-called deep geological disposal in a GDF.[13] In December 2018, the UK government launched the siting process for a GDF in England, Wales and Northern Ireland, after previous attempts to site a GDF did not result in any decisions.[14] This launch was preceded by a series of stakeholder workshops organized by Radioactive Waste Management (RWM), a UK-government subsidiary tasked with the delivery of a GDF.

The term 'disposal', referring to a permanent solution, is used in opposition to 'storage', which is considered a temporary measure. According to the UK government, geological disposal involves 'preventing radioactivity from ever reaching the surface in levels that could cause harm'.[15] But disposal some 500 metres beneath the surface also requires surface-making to keep the waste contained: a GDF is predicated on containers resisting corrosion and leakage, a suitable geology, and engineered barriers to make the enclosing geology even more secure. For operations at Sellafield, this means that engineers must anticipate requirements set out by RWM for packaging considered suitable for enclosure in a GDF. Because no decision has been made yet on where this GDF might be located, RWM requirements are based for safety reasons on worst possible scenarios for the as-yet-unknown location, which entails costlier packaging than may be required in reality. Also, with retrievals at Sellafield continuing without a permanent disposal route, additional interim stores need to be built on-site as a temporary (and costly) solution – for example, at the waste encapsulation plant, where stores are filled up. Decision-making in such matters is by no means a technoscientific issue only. As one Sellafield discussion partner sighed, 'everything is determined by money and politics'.

GDF siting process: HAW away from the surface

In the information sessions I attended in 2018 and 2019,[16] the GDF was confidently presented as the way to go: safe, secure, sensible and firmly grounded in science. The UK government hopes to locate a site for a GDF led by the principle of volunteerism. Initially, every interested party is invited to come forward, including individuals, landowners, businesses, public bodies and community groups, with promises of financial compensation.[17] It is not the geology that will be in the lead – that is, no areas are to be excluded or preferred from the outset on the basis of geological suitability. What is important to note is that, even though the principle of volunteerism implies that for now the UK government does not wish to impose a GDF on a region, and even though the approach comes across as quite open and community-orientated, the UK government does know what it seeks to achieve: to site a GDF in a geologically suitable, or suitable enough, place that is acceptable to a host population (to which financial incentives will be offered). It is, however, not uncontested. Alternatives might include keeping waste on the surface (which is the option preferred by the Scottish government) or burial in deep boreholes.[18] In other words, the government is not embarking on this process as a disinterested party testing the grounds for a nuclear waste policy – rather, it knows what it wants, and presents its chosen option with a smoothness that suggests government rhetorics as yet another form of nuclear surface-making.

West Cumbrians on a GDF – and what materials might go under the surface

Because of the region's familiarity with the nuclear,[19] and because the lion's share of the UK's HAW is already located at Sellafield – either lingering in legacy facilities or packaged and stored – West Cumbria is widely expected to show an interest in hosting a GDF. What has been ethnographically very interesting, however, is how the region positions itself strategically, in particular through the channels of the WCSSG. West Cumbria presents itself as being unique for having had stewardship over so much of the UK's nuclear waste, and demands acknowledgement for having had it on the surface for so long. It deserves more than to be a contender for a GDF – it deserves another kind of treatment. In discussions leading up to its response to the GDF consultation, members of the Enablers group (a WCSSG working group) suggested that the process proposed by RWM was not appropriate for West Cumbria because 'we are not a new host community; we have been doing this for seventy years'; 'we are the host community *now*'; 'this is a unique community in its understanding, history, impact'; 'our community has dealt with this and supported the industry for a long time'.[20] Nuclear expertise and nuclear familiarity, it turns out, are both part of West Cumbrian identity and negotiation tools. Of immediate interest to our discussion here, however, is the matter of which nuclear materials are destined to disappear permanently under the surface in a GDF, which has not yet been fully decided by the UK government. Various discussion partners in West Cumbria expressed reservations about the inclusion of plutonium in a GDF because they felt it could serve useful purposes, and they were for this reason reluctant to show an interest in the possibility of hosting a GDF.[21] The large quantities of plutonium stored at Sellafield may well be considered a security hazard, but for now the government has not taken a stance as to whether this nuclear material should be considered waste or not.[22] Some discussion partners entertained hopes for the development of feasible energy technologies that can use plutonium or nuclear waste as fuel to generate nuclear power.[23] Disturbing the neat distinction between 'on the surface' and 'deep disposal', then, plutonium plays an active part in local debates about potential candidacy for the siting of a GDF.

The geologic turn: surfaces in flux

In RWM's approach, geological disposal is presented as the unambiguous way to keep nuclear waste firmly separated from (human) life and the environment in general. A scientifically sound play of Russian nesting dolls is foreseen to keep nuclear waste away from the surface by means of containing, packaging, encapsulating, vitrifying, adding layers, engineering barriers and mimicking geology. Once a GDF is in place, RWM explains, nuclear waste will no longer be stored – it will be disposed of, put *away*.

This belief in keeping materials contained and separate appears to be in tension with a recent scholarly turn that insists on intimate connections between human lives, materials and geologic scales, and which problematizes the common assumption that materials are passive, awaiting human imprint, instead conceptualizing these as 'vibrant'. In *Making the Geologic Now: Responses to Material Conditions of Contemporary Life* (2013), editors Ellsworth and Kruse have gathered case studies and practices that emphasize connections between human and geologic timescales, arguing that this is not only analytically productive but also necessary from a moral perspective for keeping the planet alive. They seek to trouble 'categorical distinctions between the "brute materiality" of geology's "external world" (rocks, minerals, mountains) and the soft, "inner" worlds of biology's living things', arguing that humans do not observe the geologic as a landscape, but instead 'inhabit the geologic' (Ellsworth and Kruse 2013: 17, 25).

One of the volume's case studies offers interesting thoughts on deep geological disposal. By means of an interview with geoscientist Abraham Van Luik, Manaugh and Twilley (2013) provide technical information on the behaviour of potential host materials for a GDF (clay, granite, salt) and their interaction with packaging materials, focusing on the case of Yucca Mountain in the United States, a site destined for a GDF yet remaining in legislative limbo.[24] Particularly interesting are examples of scientists looking for 'natural analogies', mimicking conditions in computer simulations, in a bid to get engineered barriers right. Van Luik discusses 'natural reactor zones' in the Oklo mining district in Gabon, where ore bodies that were highly radioactive two billion years ago might provide an analogy for the behaviour of buried HAW in the future (cf. Hecht 2018). The ore bodies, he explains, 'have been subjected to oxidizing conditions, because uplift of the land brought them above the water table'. The implication, then, is that geologic surface-making led to the formation of these scientifically interesting zones.

In these approaches inspired by the geologic turn, the emphasis is firmly on affordances of combining, connecting, making analogies, both conceptually and in practice. Surfaces, in these approaches, are in constant flux. Conceiving of surfaces as ever-expanding layers of sedimentation accruing experience and time, Giuliana Bruno relates surfaces to change, and change to alchemy. 'The physicality of a thing one can touch', she suggests, 'does not vanish with the disappearance of its material but can morph culturally, transmuting into another medium. I like to call this technological alchemy' (Bruno 2014: 7).[25] Morphing appears to lend a material or thing a life of its own, and this ties in with an insight that has emerged from studies embracing thing theory and materiality, namely, the insight that matter must be seen not as a passive receptacle for human action, an amorphous mass that awaits intelligent imprinting, but as an entity that may not possess the intentionality of agency yet plays an active role in more-than-human relationships whose relationality goes beyond a one-way street (e.g. Barad 2003; Ingold 2010, 2011: 178; Gregson 2012; Harvey 2019). From this perspective, matter may be considered 'vibrant' (Bennett 2010).

The vibrancy of matter is central in Gregson's (2012) plea for an acknowledgement of the thing power of nuclear waste, that is, its power to disrupt. She suggests this acknowledgement be made by forging and maintaining attachments with HAW, particularly when it dwells deep down under the surface, while being aware of its threat, rather than seeking to dominate and tame it from a human-centred technoscientific perspective (ibid.: 2008). Gregson convincingly argues that there has been a *discursive* taming of HAW, presented in tables and diagrams to the public, in contrast with the more visceral experience she had at Sellafield during visits in 2007–9:

> Current NDA public displays and simulations of a DGF [Deep Geological Facility] do not admit a sense of thing power. Rather, as befits the discourses through which they are thought, they are about the technical mastery of HAW. Within experimental work, however, the energetic capacities of HAW, specifically its distributed potential effects within a DGF, are well to the fore, albeit framed within technical discourses of control.
>
> (Ibid.: 2017)

These discourses of control, however, appear to relax in more informal contexts. Ethnographic work lends itself to conversations that delve beneath a well-maintained surface of corporate confidence, opening up spaces that allow doubts and uncertainties to be voiced. This, after all, is key to the practice of nuclear decommissioning. Retrieval of nuclear waste stands in tension with the goal of containment, of sealing under a protective surface, because it requires a temporary porosity. This is the paradox of nuclear decommissioning. I suggest that the two materials touched upon above help to complicate smooth surface narratives: sludge as it disturbs the story of controlled retrieval from the legacy pond, and plutonium as a contested candidate for deep geological disposal.[26] In formal, overly positive discourse about technoscientific mastery of nuclear waste management, these materials are presented as challenges to safety and security for which solutions are pursued. In discussing sludge or plutonium in more intimate settings, the message of control is abandoned for a richer conversation that acknowledges the material's vibrancy or potential – and how problematic this may be when one is trying to decommission a nuclear facility safely and securely. Nuclear historian Gabrielle Hecht, reacting to the recent move to call materials 'alive' or 'vibrant', has rejected the epithet 'vibrant' for waste because of the danger of fetishizing it, lending it a mystical quality as if it cannot be known, which she considers a politically questionable move (2018: 112). I suggest, drawing on conversations with those that have close encounters with such materials, that embracing sludge as an enigma and plutonium as an ambivalent, historically and politically charged substance might open up avenues for public discourse on nuclear decommissioning that allow uncertainties to be voiced and acknowledged, rather than smoothed over and kept under the surface, only to fester.

Notes

1 My fieldwork at and around the Sellafield nuclear site began in August 2017 and is ongoing, with Sellafield Ltd (SL), the company operating the site, having kindly provided me with access. Special thanks are due to SL's Andrew Cooney for his patient and astute advice and for greatly facilitating my ethnographic work. The ethnography focuses on the transformations taking place at SL and in West Cumbria during a period when the nuclear site is moving into full decommissioning.

2 The distancing epithet 'legacy' refers to nuclear facilities and wastes that stem from a previous period of nuclear production and have been 'inherited' to be dealt with by the current generation.

3 For the differences between the various waste categories (which are a matter of characterization), see https://ukinventory.nda.gov.uk/about-radioactive-waste/what-is-radioactivity/what-are-the-main-waste-categories/ (accessed 15 August 2019). Vitrification is a method of immobilizing HLW by incorporating it into a glass matrix. Intermediate- and low-level waste may be immobilized through encapsulation in containers, mixed with grout.

4 For the UK's National Nuclear Laboratory, see https://www.nnl.co.uk (accessed 15 August 2019).

5 Several facilities have been built elsewhere. Perhaps most intriguing among these is the Onkalo waste repository in Finland (not yet in use), portrayed in Michael Madsen's documentary *Into Eternity* (2010).

6 See https://www.gov.uk/government/organisations/nuclear-decommissioning-authority (accessed 15 August 2019).

7 See, for example, the story of a robotic arm being deployed in the Magnox Swarf Storage Silo (Sellafield Limited 2017: 3); or the hole cut into the Pile Fuel Cladding Silo (PFCS), 'a locked vault which was never designed to be opened', see https://www.gov.uk/government/news/first-hole-is-cut-in-worlds-oldest-nuclear-store (accessed 15 August 2019). On the PFCS as an 'intolerable risk' and the Office for Nuclear Regulation agreement on how to tackle it, see the following report: http://www.onr.org.uk/pars/2018/sellafield-18-023.pdf (accessed 15 August 2019).

8 'Value stream' is the official term for SL's main organizational units.

9 This is the formal body representing West Cumbria in its public scrutiny of the nuclear industry, set up in its current format by the NDA; see www.wcssg.co.uk (accessed 15 August 2019). It has several working groups that meet at regular intervals in the course of a year, and the public are welcome to attend. Similar scrutiny groups exist in other UK regions where nuclear decommissioning takes place.

10 The NDA included the following entry on its website on de-sludging in 2016: 'Radioactive sludge has been transferred out of the world's oldest nuclear fuel pond for the very first time, marking a historic achievement in the decades-long programme to clean up Sellafield. … The project is being delivered 10 years ahead of schedule and for half the expected cost.' See https://www.gov.uk/government/news/radioactive-sludge-removed-from-sellafield-pond-for-first-time (accessed 15 August 2019). Cf. a recent news update on a make-do solution for PFSP: https://www.gov.uk/government/news/pond-cleaner-ready-to-make-a-splash (accessed 15 August 2019).

11 A lot of effort goes into more sophisticated waste characterization, resulting in possibilities for the diversion of very LLW outside a nuclear licensed site and, for example, the recycling of metals: see https://www.gov.uk/government/news/llwr-and-sellafield-contracts-help-cyclife-thrive (accessed 15 August 2019).

12 'The term Higher Activity Waste (HAW) refers to all radioactive material that has no further use that falls into the following categories: High Level Waste (HLW), Intermediate Level Waste (ILW) and the relatively small volume of Low Level Waste (LLW) that is not deemed suitable for disposal at the LLWR or the LLW facility at Dounreay [in Caithness, Scotland].' See https://www.gov.uk/government/publications/nda-higher-activity-waste-strategy (accessed 15 August 2019).

13 International initiatives to site GDFs were discussed at the Integration Group for the Safety Case symposium 'Current Understanding and Future Direction for the Geological Disposal of Radioactive Waste' that I attended in October 2018 (see http://www.oecd-nea.org/rwm/workshops/igsc2018/ (accessed 15 August 2019)), co-organized by the International Atomic Energy Agency and the European Commission.

14 For a history of the UK government's attempts at siting a GDF, see Gregson (2012) and Blowers (2017). For the UK's current approach, see RWM's website: https://www.gov.uk/government/organisations/radioactive-waste-management (accessed 15 August 2019).

15 See www.gov.uk/guidance/why-underground (accessed 15 August 2019).

16 Meetings that I attended on the GDF siting process included a pre-consultation workshop in Lancaster on 20 March 2018; a WCSSG Enablers meeting in Egremont on 12 April 2018; an RWM consultation in Penrith on 20 February 2019; and an NDA and RWM regional engagement event in Whitehaven on 18 September 2019.

17 See https://geologicaldisposal.campaign.gov.uk (accessed 15 August 2019).

18 These possibilities were mentioned in an interview I had with a former Sellafield deputy head. The GDF consultation just pushes its one solution, he sighed, 'just like Theresa May' (our conversation took place a week before the date that the UK was originally supposed to leave the European Union in March 2019). Also, critics of nuclear energy see a GDF as opening the way to produce more nuclear waste without thinking properly about ways to avoid its production. See for example my interview with activist David Lowry: https://www.mub.eps.manchester.ac.uk/thebeam/2019/02/18/talking-waste-disposal-with-david-lowry (accessed 15 August 2019).

19 At a recent workshop I co-organized on future-making post-decommissioning in West Cumbria ('Sellafield Site Futures', www.mub.eps.manchester.ac.uk/thebeam/2019/10/11/exploring-sellafields-limitless-future/ (accessed 15 November 2019)), one participant working in the nuclear industry remarked that West Cumbria's most valuable asset (for future-making) was the tolerance that local people have of the nuclear.

20 For more on a GDF from a West Cumbrian perspective, see Kalshoven (2020).

21 As in an exchange with a Sellafield union member: 'Are you in favour of a GDF in West Cumbria?' I asked. 'No', he answered to my surprise. 'I first want to know what goes into it. Not what can still be used! That should be retrievable.' He suggested that the physical size of the GDF would also depend heavily on what was meant to go in: 'Last time [during a previous consultation]', he said, 'the size of the area that was proposed was frightening! We *do* need to put distance between that which is really useless and humans – on the other hand, it can't be that dangerous if I am working next to it, can it?' 'Well', I said, 'I suppose there are security concerns as well …' He acknowledged this as we looked out over the site from an elevated vantage point. The way to go, he suggested, might be a small GDF for the *really* high-level waste, and near-surface storage for the rest.

22 For the NDA's approach to plutonium, see https://assets.publishing.service.gov.uk/government/uploads/system/uploads/attachment_data/file/791046/Progress_on_Plutonium.pdf (accessed 15 August 2019).

23 As in so-called fast breeder reactors, with which the UK has had experience at the Dounreay research reactor, which is now being decommissioned. See https://www.gov.uk/government/organisations/dounreay (accessed 15 August 2019). For a strong rejection of the 'plutonium economy' and a passionate plea against breeder reactors, see Von Hippel, Takubo and Kang (2019).

24 For reflections on Yucca Mountain and the concept of deep time, see Bloomfield and Vurdubakis (2005) and Ialenti (2014); cf. Irvine (2014) on human engagement with deep time.

25 This approach is also evident in *Cumbrian Alchemy* (2013), in which professor of art Robert Williams and artist Bryan McGovern Wilson make connections through time (geologic, mythical, Neolithic, human) by focusing on vibrant materials in the energy industries and on archaeological sites along the West Cumbrian coast. They conceive of this exploration as a practice of alchemy. Wilson also contributed a piece to *Making the Geologic Now* (2013) in which he takes on the persona of Robert Oppenheimer roaming the Trinity nuclear test site in New Mexico, experimenting with a process of analogy through re-enactment.

26 They also complicate things because of their sheer materiality. Both materials are interesting from the perspective of powder morphology, which is all about the multiplication of surface – the finer the powder, the more surface it possesses. Plutonium is generally stored in powder form, which makes it dangerously respirable. Sludge, once disturbed, increases in surface, and when dispersed it becomes more difficult to handle. I am grateful to my colleague Professor Richard Taylor for making me aware of powder morphology and its association with surface.

Chapter 3

Surface Eruption: Machine Creativity and Emotive Data Objects

Barbara Rauch

This chapter addresses the transformation of surface qualities in artistic printmaking practice, focusing on the process of digital printing onto paper and the 3D printing of material objects. It critically examines the qualities that may be gained or lost as the objects move from digital instantiation – usually in a database or on a computer screen – to its physicalization into 2D- or 3D-printed artefacts. Through the examples of my own practice and of others', this chapter discusses the multiple processes at play when translating synthetic digital data into objects with material surface qualities, to which the human senses can relate. It offers, from a creative practitioner's point of view, the potentials of machine creativity and artificial emotions by attending to the human maker, synthetic material and their agencies.

Many artist studios have now embraced digital printing, laser cutting, haptic sculpting, 3D prototyping and other means of data materialization techniques, creating surfaces which are no longer reliant on physical manipulation by the human hand. They rather materialize the technologies that have created them. But how do these technologies influence the reception of artistic practice and the aesthetics of artwork? Many contemporary new media artists are working in a post-digital position that no longer foregrounds digital technology in their work. As the digital becomes embedded in artistic practice to the point that we actually 'think with it' (Agamben 2002), the distinction between the digital and the non-digital is no longer useful. Accordingly, from a post-digital position, this chapter critically questions how we think, feel and create with and through the digital, and what it means to be human in contemporary technological conditions.

In the early 1990s, when the distinction between a physical surface and a virtual version of it seemed much clearer than it is now, philosopher and media theorist Vilém Flusser (1991) explained his unique understanding of a technical image as the coming together of surface imagination and programmability. Flusser seems to propose that new technology can bring human imagination and digital computation together. The algorithm is a combined effort between human and computer, as it translates an abstract human vision into something more concrete; that is, as a visualization on a computer monitor or as a materialization in print. An apt example of this reading is Flusser's own fifteen-year collaboration with the artist and scientist Louis Bec. Their work *Vampyroteuthis Infernalis* (Flusser and Bec [1987] 2012) is an early example of such synthetic speculation. In this work, Bec's drawings illustrate fictitious creatures that reflect different aspects of Flusser's philosophical thinking. The creatures are presented as zoosystemic plates and appear to be hybrids of natural and artificial species. Artist Eduardo Kac sees the work as a cutting-edge experiment into animal cognition, philosophy and art (cited in ibid.). It is this potential of synthetic worlds that drives artist Georg Mühleck's practice. Mühleck's print works – usually on mural-sized paper – are generated by algorithms that he employs, which are mathematical formulae that mimic natural life. As one of the algorithms employed for his work about artificial life is called 'Life' – this is one of the better-known rules of cellular automata – we cannot help but compare these digital iterations of synthetic data with the evolution of natural life.[1] The digital algorithm remains in an in-between state that can offer an infinite number of potentials. Comparable to Flusser and Bec's synthetic speculation, Mühleck's work is the result of the artist–algorithm combination, and the theme, fittingly, is about genetic manipulation and imaginary habitats. Mühleck (2019) describes a series entitled *kelpies* (Figure 3.1) as 'mind creatures' which are a 'symbiosis of plant, animal and human being'. Similarly, the artist's statement for another series entitled *caspecies* (Figures 3.2 and 3.3) illuminates his intentions:

> The word 'CASPECIES' consists of Cellular Automata and SPECIES. These species are organisms comprised of Cellular Automata. They nourish on algorithms. In the process of creation, cell cultures go through hundreds to thousands of generations before they freeze into a large artificial micro-still. Rather than simulating real life, caspecies evoke the possibility of creatures of yet unknown origin and scale.
>
> (Mühleck 2019)

By creating these speculative objects – *caspecies* or *kelpies* – Mühleck seems to unsettle existing boundaries between human-creator and manmade machine.

Figure 3.1 *Inked kelpies*, *c.* 2018. Archival pigment on hot-press natural fine art paper 300g, 108 × 77 cm (each), edition of four. © Georg Mühleck.

Figure 3.2 *Staged caspecies* (from the first and the second group), *c.* 2015. Archival pigment on Canson 310g, 108 × 77 cm (each), edition of four. © Georg Mühleck (see p. 54).

Figure 3.3 *Staged caspecies 12* (detail), *c.* 2015. Archival pigment on Canson 310g. © Georg Mühleck (see p. 55).

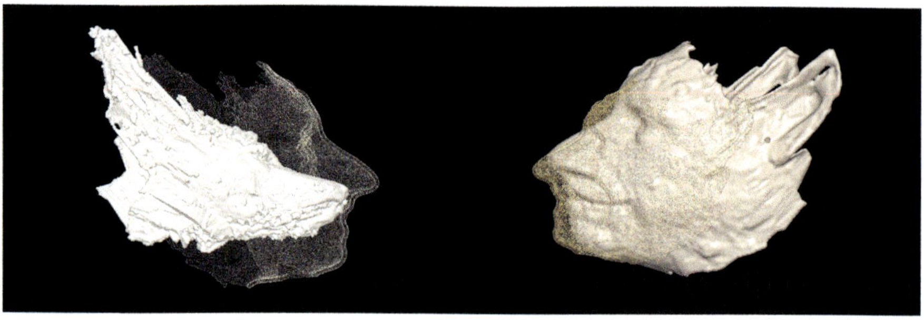

Figure 3.4 *interFaced* (screenshots), *c.* 2009. Geomagic FreeForm (3D rendering software). © Barbara Rauch.

Cellular automata rules are abstracted processual simulations employed in several scientific disciplines – here in Mühleck's work, they are uniquely used to create imaginary hybrid life forms. Although the creatures only exist in an artistic context, some synthetic biology lab results do not look too dissimilar to the examples presented here.

This transgression of boundaries inspires my own practice. For example, the series entitled *interFaced* (2007–13) (Figure 3.4) calls for complex questions regarding our interaction with digital or material objects, as well as the location of authorship and creativity. This series derives from a large database of digitized faces. Although the 3D-printed outcome looks like a collection of hand-sculpted objects, they digitally evolved without any manipulation by human hands. For the series, I collaborated with scientist Peter Hammond and his team (Institute of Child Health, University College London), who provided me with a large database of human faces with neutral expressions, and assisted with combining this database with another set of data, this time with distinct facial expressions. In addition, some animal scans were also combined. The resulting hybrid faces presented animal and human features with recognizable expressions. Hammond and his team specifically designed an algorithm that allowed animation of the 3D data objects, which led to a series of large-scale animated projections and 3D-printed objects. The results came to me as a surprise. With the database objects, the data dictates the form while the technology negotiates with the limitations of the hardware and software. There is little human involvement except for responding to potential glitches and mistakes caused either by hardware or software, or the artist's limited programming skills. The project seems to show that artificial intelligence can allow for the creation of artworks, which would not be possible otherwise. Before long, artificial intelligence might gain artistic autonomy, as machine learning, neural network training and pattern recognition all aim to mimic the capacity of the human brain. The series *interFaced* addresses the process in which machine intelligence, material affordance and affective data materialization interact with the viewers, opening up questions about creativity and authorship in post-digital artistic practice.

A related example was displayed at the *3DXL* exhibition in Toronto (Design Exchange, 14 May–16 August 2015). The exhibition demonstrated the latest 3D-printing applications within architecture. The work that the architects Dillenburger and Hansmeyer presented, the *Arabesque Wall*, is of 3D-printed sandstone of almost twice human height. The architectural component displayed a complex geometry with fine details, demonstrating the potential of digital fabrication that could not have been realized using traditional methods. The architects confirm that the algorithm, rather than their imaginations, had produced the wall: 'The algorithms are deterministic, as they do not incorporate randomness, but the results are not necessarily entirely foreseeable. Instead, they have the power to surprise' (Hansmeyer 2013).

Considered together, *interFaced* and *Arabesque Wall* call for another question. Most of the materialization process – that is, the physical engagement with the material (sandstone, ceramic powder or polymers) – was done by a machine. The objects were 3D-printed in layers, in a manner very different from how a stonemason or sculptor would have engaged with the material. As the boundaries between the analogue and the digital become increasingly blurred, machine- and hand-made objects now occupy a controversial yet shared space.

When asked about what a work may gain or lose as it leaves a computer monitor for a paper substrate, Mühleck remarked that the digital file is animated with potential, while the physical print on paper is a 'still life', representing only a slice of the possibilities possessed by the digital file. On the other hand, Mühleck also admits that working on a digital file does not involve his own dynamic responses, and that he finds this 'removed' work process frustrating. It is for this reason that he usually extends the work to a printout, making specific choices of paper substrates and the qualities of wide-format printing. Working on a tangible material meets the artist's desire to hold the work in his hands. The work comes across quite differently when the 'digital creature' (Mühleck's term) is transported onto the paper surface, as it seems to gain a certain haptic quality. The printed creature looks as if it is coming alive, to the point that the viewer may be triggered to touch and lift it off the printout. For the series *interFaced*, I experienced the satisfaction of handling virtual objects by further manipulating the surface of the objects with a haptic sculpting tool. The Sensable FreeForm system coupled with the PHANTOM haptic device allowed me to manipulate the objects on the monitor, and to feel the physical feedback on my fingertips. The sculpting pen can be coded for different material densities: soft or hard clay, glass or wood, for example. Such 'virtually physical' interaction prompted me to produce printed objects that I could actually handle. This process involved the consideration of many variables such as printing technologies and the quality, dimension and availability of materials.

Algorithmic creativity surprised me again in another example of my work. For *Thick Friendship* (2013) (Figure 3.5), busts of Goethe and of Schiller were 3D-scanned and morphed together with a 3D scan of a deer skull. This work refers to the long friendship between the two prominent figures and Goethe's

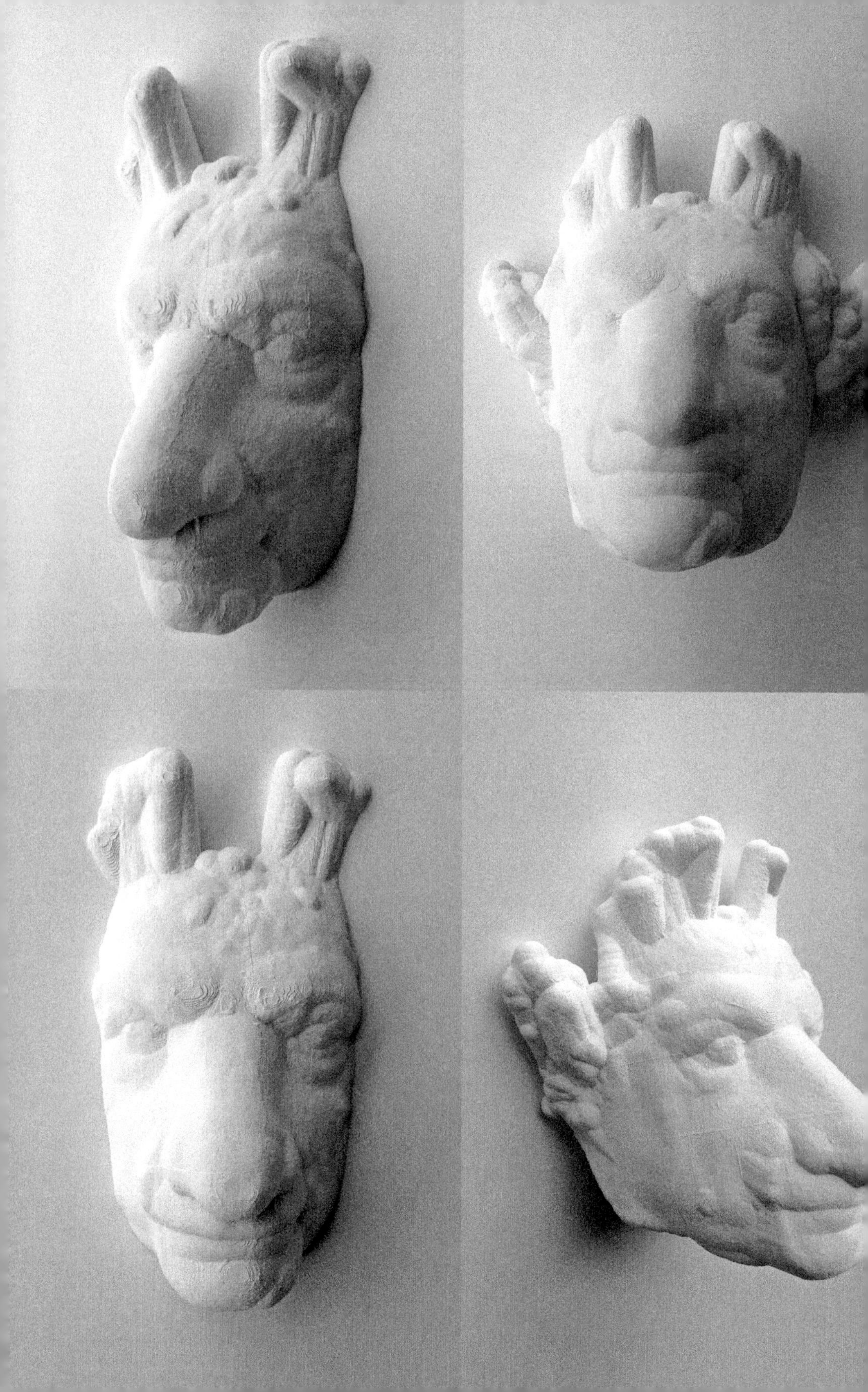

recurring dream about a skull, about which I read in Rüdiger Safranski's *Goethe und Schiller* (2009). The physical objects are made of laminated laser-cut paper and produced using Laminated Object Manufacturing (LOM) technology. As was the case with *interFaced*, the aesthetic qualities of *Thick Friendship* are mostly an algorithmic result. As most of the process is hidden in the complexity of the algorithm, I find it nearly impossible to foresee the final results. The work, therefore, seems to highlight the aspects of emotion in machinic creations. If artificial intelligence allows for the creation of artworks with aesthetic qualities indistinguishable from those of human creators, will they be able to mimic human emotions too? In the New Aesthetic discussion, digital 'eruption' (Sterling 2011) refers to the manifestations of the digital – the styles of computer games or pixelated images – within a physical setting. In my work, of which the aesthetic quality is distinctively human, this 'eruption' becomes somewhat literal, as the digital information 'breaks through' the surface of the object, presenting us with folds and bumps rather than a smooth surface (Figure 3.6). Data materialization or re-materialization thus articulates the complexity of the digital materiality. As a fusion of the material and the immaterial, it introduces new and unanticipated objects. As the examples above show, my practice encourages the non-human agency of the artificial to create its own synthetic surface. Our new tools can allow for manifestations that were previously considered impossible, leaving human imagination a step behind.

Flusser ([1985] 2011: 13) states that the technical image does not represent a world that is already out there, but rather provides a window through which one recognizes a universe of computation. Technical images have since become ubiquitous, to the point that art critic Claire Bishop (2012: 436) suggests that, while inconspicuous in the mainstream art world, 'the digital is, on a deep level, the shaping condition – even the structuring paradox – that determines artistic decisions to work with certain formats and media'. My work, to some extent, attempts to respond to Bishop's (2012) question: 'While many artists use digital technology, how many really confront the question of what it means to think, see, and filter affect through the digital?' Out of the impact digital technology may have on many aspects of contemporary human conditions, I am particularly interested in the aspects of creativity, emotions and affect. With the digital and physical examples of artist–algorithm collaborations, which depict the transgression of human–animal–plant and/or natural–synthetic boundaries, this chapter explored new printed surface qualities that 'erupt' as the digital manifests itself as a material surface. The tactility of the new surfaces offers a critical insight into how we understand a material object and the technology that made it come to life.

Figure 3.5 *Thick Friendship*, (opposite) *c.* 2013. Paper, 28 × 19 × 19 cm (each). © Barbara Rauch.

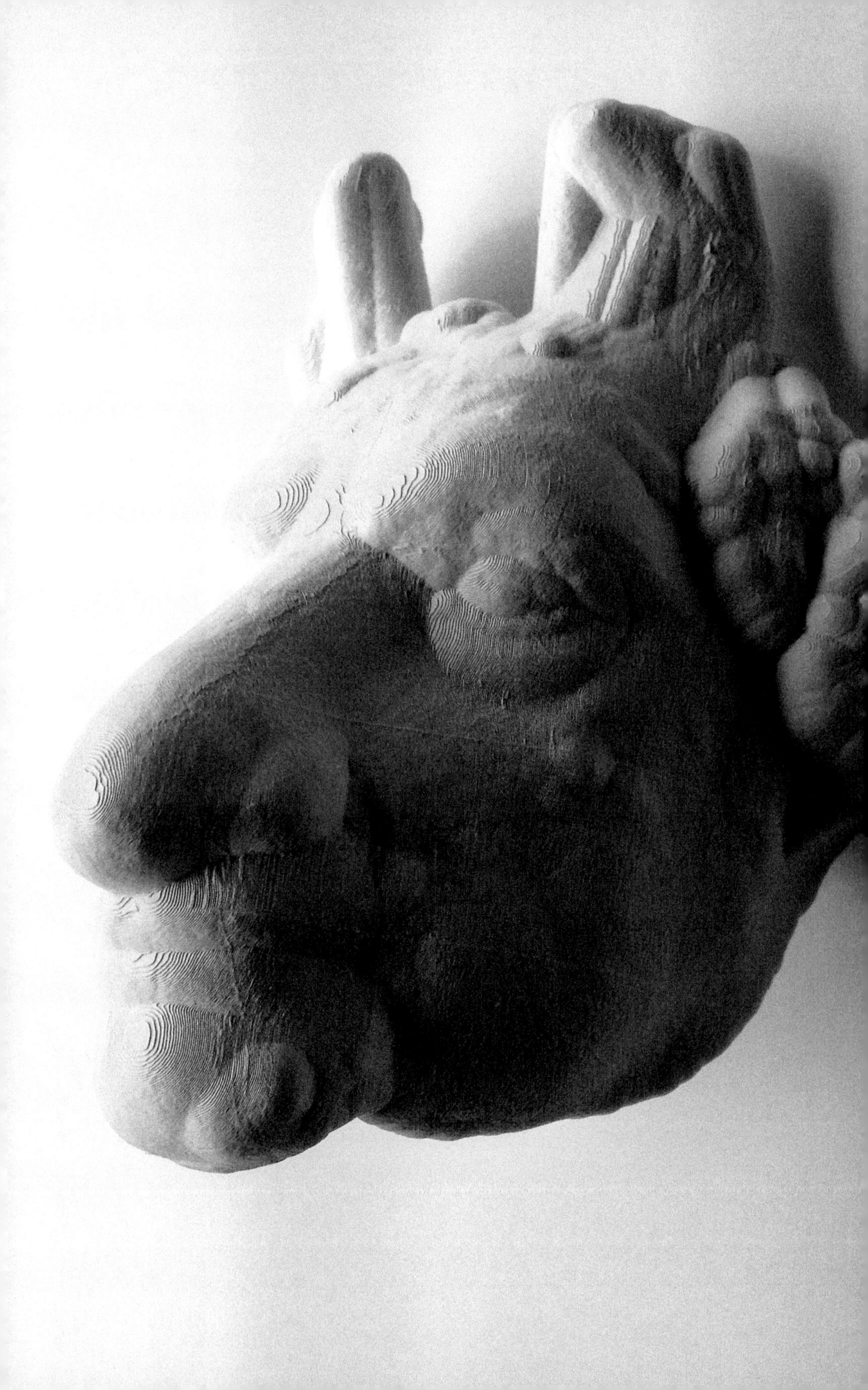

Note

1 'Life' or 'The Game of Life' is a cellular automaton discovered by John Horton Conway. While using simple rules, cellular automata can simulate a variety of complex real-world scenarios. In simple terms, it uses Von Neumann's self-replication system to simulate evolution. Other rules used by Mühleck include: 'Brain' (Brian Silverman), 'The Hodgepodge Machine' (Heike Schuster and Martin Gerhart), 'Diffusion' (Tommaso Toffoli and Norman Margolus), 'Growth' (A. K. Dewdney, Leonard M. Sander) and 'Ranch' and 'Faders' (Rudy Rucker). In this context, Mühleck would like to credit Ken Karakotsios for programming an accessible interface.

List of Figures

Figure 3.6 *Thick Friendship* (detail), opposite, *c.* 2013. Paper. © Barbara Rauch.

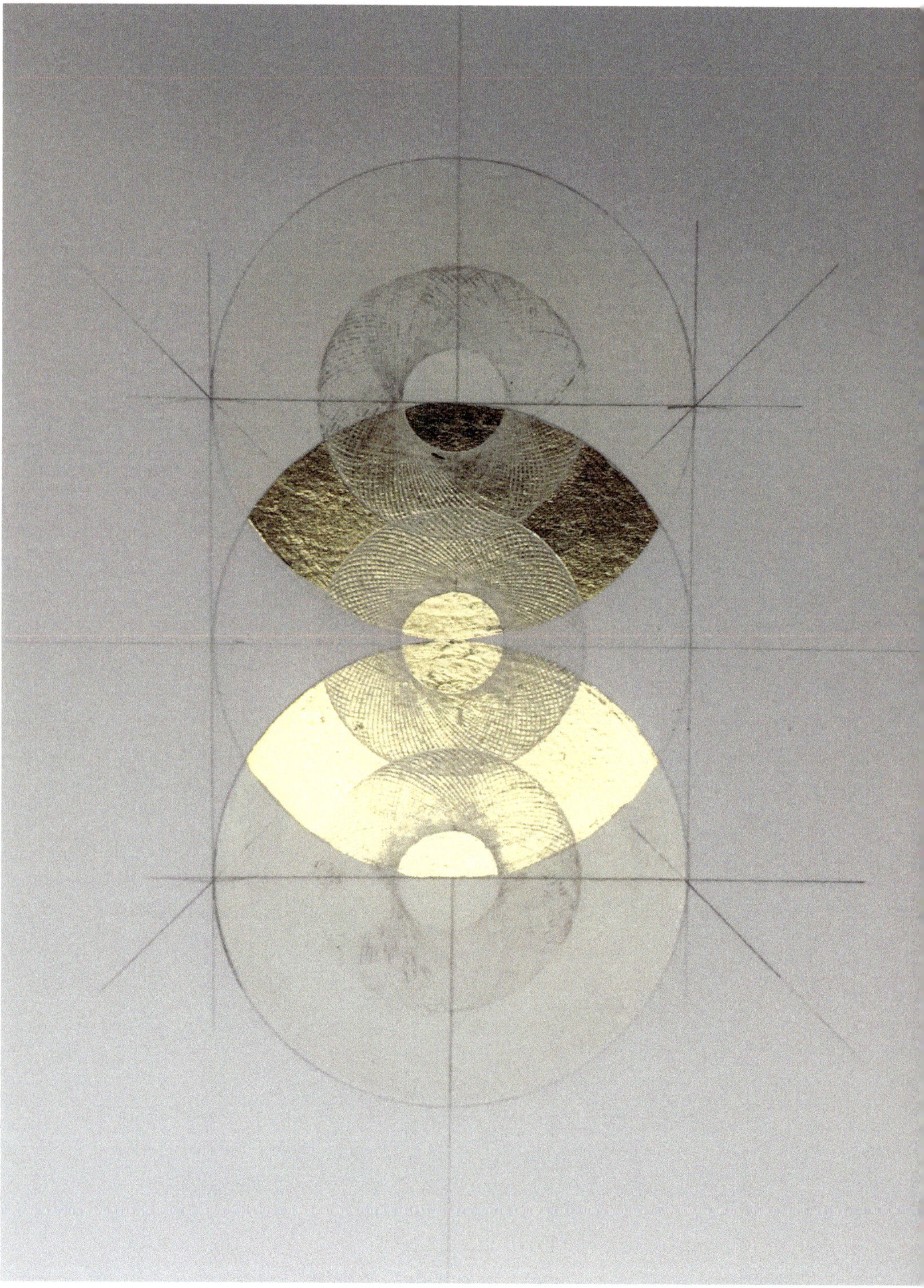

Chapter 4
The Depth of Surface

Lesley Halliwell

My visual art practice explores interrelationships between surface, pattern, distance and time. Using my creative practice as a process of investigation, here I probe where the surface of an artwork resides. As I consider the interplay between the outward and inward components of the picture plane, I am moved towards an understanding of surface beyond the dichotomies of seen–unseen or inner–outer. Through a series of reflections in response to specific artworks – forgoing the conventional essay form – this chapter offers multiple ways of thinking about surface and its complexities.

Figure 4.1 *Tilted Plane*, c. 2015. Pencil, lacquer and silver leaf on paper, 25 × 35 cm. © Lesley Halliwell. What is the interplay between the outward-facing components of the picture plane – what we can see, feel and touch – and the inward elements that support or modify what we see – such as a preparatory design, traces of working process, or a generating framework? My interest in surface arose from a sustained preoccupation with the relationship between pattern and structure, which remains a foundational pursuit within my artistic practice. Surface – described by Christa Robbins (2002) as 'a vacillating plane, receding and materializing throughout the history of art, often at the centre of some of its most ideological and teleological concerns' – is a pivotal feature in fine art practice. The varied aspects of pictorial surface are epitomized in, for example, Renaissance perspective, *trompe l'œil* realism, colour field painting or unvarnished canvases. Still, it is common to conceptualize surface as a superficial coating on the outside of the picture plane. But it is more like a skin that can both conceal and reveal things that take place beneath. Moreover, there is the area between that multivalent skin and the viewer, where effects of light and reflection can radically alter what is perceived. In line with anthropologist Tim Ingold's (2017: 100) challenge to the cultural assumption that 'truth can never be on the surface but [lies] somewhere deeper down', my practice explores the depth of surface to reveal that relations of above–below or outer–inner are fluid, so we might better think of zones of *interactivity*.

Figure 4.2 *Fanatic, 4500 Minutes* (detail), *c.* 2011. Ballpoint pen on paper. © Lesley Halliwell. A child's plastic drawing tool, the Spirograph, is used to repeat the same motif, step by step, incrementally building the image during an extended period of time. These drawings are titled by the total of the segments of time they took to make, referencing the clocking-on and clocking-off of the workplace, while the ballpoint pen leaves its trail of ink across huge planes of white paper. On closer inspection, the 'handmadeness' of the drawing is as apparent as the mechanical: concentration wavers, hands slip, pens run dry, paper wears out. These imperfections – the remnants of actions that counteract the mechanistic aspects of the process – remain visible on the paper, thus revealing the surface structure and its fragility.

Figure 4.3 *G(u)ilty* (detail), *c.* 2016. Gold leaf, lacquer and pencil on paper. © Lesley Halliwell (see pages 66–67). Reflecting on my practice, I noticed how actions that eat into the surface, such as indenting, eroding or burrowing in, recur irrespective of the media being used. In *G(u)ilty*, a series of intimately scaled drawings, the metal nib of the ballpoint pen indents the gilded surface, playing with the outwardness and inwardness of the pictorial plane. The drawing process involves adding a layer of gilded surface onto the existing surface and acting upon those layers, rather than constructing the drawn surface from ink alone. While pattern focuses the eye on the surface, flattening the plane in such a way that it hinders entry, actions that interfere with the pattern create an 'entry point' for the viewer. In this case, the line that scores the layers stirs up the surface and confirms its materiality. This allows the gaze to discern, and the mind to understand, the complexity of the surface in a new way.

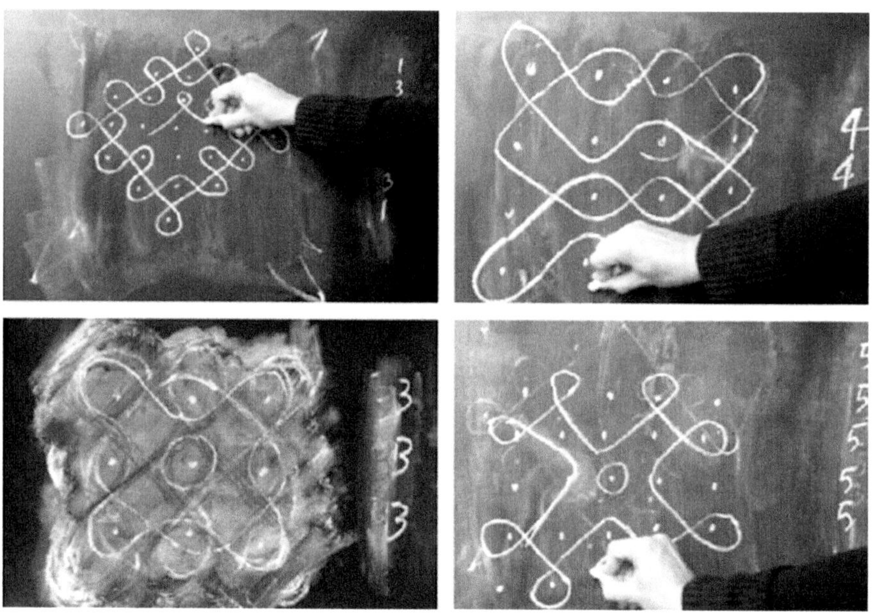

Figure 4.4 *Additive Trace* (film stills), *c.* 2015. © Lesley Halliwell (see page 68). *Kōlams* act as threshold signs of the home or temple, protecting those within from malevolent forces (Mall 2007). Practiced by Hindu women of the southern Indian state of Tamil Nadu, they are traditionally made by trickling a line of rice flour or chalk dust through the fingers. Drawn on or around a grid of pre-planned dots, a *kōlam* usually appears to be a single continuous line. As the day wears on, it is rubbed out by the passage of feet, to be renewed the following morning with a fresh pattern (Gell 1998: 85). Initially I had wanted to understand how these complex patterns were constructed without relying on the ruler or compass, and the film *Additive Trace* exposes both my drawing process and my learning process. Drawn on a blackboard with chalk, each completed image was wiped away before the next could begin, to assimilate the process with that of *kōlam*-drawing. A layering effect begins to occur as patterns are repeatedly rubbed out and worked over: an apparition-like palimpsest of my movement. There were moments when I felt 'at one' with the drawing process, a glimpse of how it might be for the women of Tamil Nadu with their free-flowing lines and loops. The anthropologist Alfred Gell compares such rhythms of movement to dancing, with the drawn image 'a kind of frozen residue left by this manual ballet' (1998: 95) – in both cases, a series of gestures embeds itself within the body over time, requiring less and less conscious thought. Eventually, the internalized rhythm surfaces. However, more importantly, *Additive Trace* also exposes my mistakes and hesitations as the chalk squeaks, scrapes, snaps clumsily over the board. I stop and pause, thinking about where to position the chalk, in which direction to take the line and how to navigate the dots. This is what Paul Virilio refers to as an 'interruption': a break that inserts uncertainty into a constructed system (Lotringer and Virilio 2005). This break is what I consider an entry point, a pause in the pattern or the rhythm that allows the activity of looking to reach beyond the picture plane. Meanwhile, the palimpsest introduces artist Richard Galpin's (1998) idea of erasure as part of the layering process: erasure creates drawing, and drawing creates erasure. In these ways, sequence is eclipsed, time is thickened, and surface is volumized, creating multiple temporalities and spatialities.

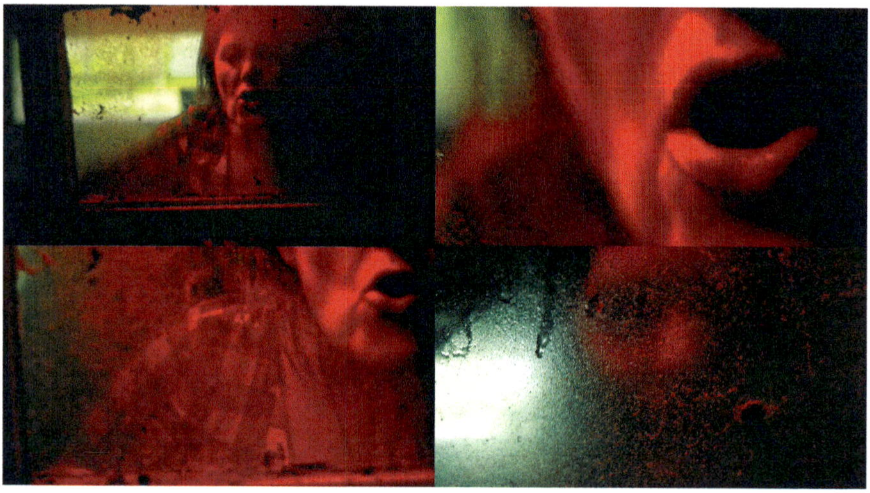

Figure 4.5 *Drawing Breath* (film stills), *c.* 2017. © Lesley Halliwell and Pavel Prokopic.

I had been thinking about using the condensation of the breath to create a temporary drawing surface on a sheet of glass. While I was taking part in a Sidney Nolan Trust residency in 2017, an opportunity arose to collaborate with filmmaker Pavel Prokopic. Rummaging in one of Nolan's barns filled with abandoned household objects from old prams to bicycles, we came across a wooden window frame with the glass caked in layers of dust. Using it as a base, we worked in the barn with shafts of light entering beneath doors and between broken roof tiles, birds flitting in and out from their nest in the rafters. … I exhale … long and slow … an elongated moment. … Condensation, mixed with the years of dirt, fills the pane, thwarting our ability to look *through* the glass. Instead, our gaze is arrested *at* the material surface (Brown 2001: 2). Like Giuliana Bruno's (2014: 231) description of the surface 'as a form of contact that enables intimacy', here is a surface that emerges with the stains of time and an intimate exchange of warm breath. In hindsight, I had initially overlooked the significance of the in-breath, that support mechanism and underlying structure: breathe in through your nose, out through your mouth. Inhale (prepare) – exhale (action). They support each other, as the inward and outward components of breath described by Peter Schwenger (2006: 81) do: 'I hear a rising, then a falling, in and out. A greatness, a lightness. I grow heavier, and then so inert that my body seems without life. Between breaths, I lose feeling. And then my chest fills, a resurrection.' Between the two movements of breathing, there is also a liminal space, a slight pause, an entry point.

Figure 4.6 *Stasis* (detail), *c.* 2015. Ballpoint pen and gold leaf on paper. © Lesley Halliwell. Interventions that allow entry can be contrasted with qualities that appear to mask the penetrability of surface, notably shine and reflectivity. In *Stasis,* gold leaf was used in the central area of the drawing. The gilding appears to 'hold' the eightfold rosette in place, preventing any sense of rotational movement. At the same time, the ability of the gold to reflect light disrupts the flatness of the surface by pushing it out towards the viewer. The metal leaf both reflects and absorbs light, allowing progression or recession. This perceptual experience led me to think about the space between the viewer and the physical surface of the work as a plane of activity. Perhaps it is this 'surfaceness' that allows the artwork and viewer to come together.

Figure 4.7 *Portal*, *c.* 2000 and 2017. Ballpoint pen and silver leaf on paper, 60 × 70 cm. © Lesley Halliwell. The black Spirograph square of the background in *Portal* was drawn in 2000. It existed as a work in its own right, until I revisited it over a decade later and 'made strange' a surface I thought I knew (Berman 2015). On top of this contained and bounded shape now sits a gilded porthole (or is it a mirror?) that opens up the present, past and future. But the tear in the metal leaf ruptures the picture plane, reminding us of the surface materiality. Nothing is stable here. Over time, ink will fade, and silver leaf will oxidize, tarnish and darken.

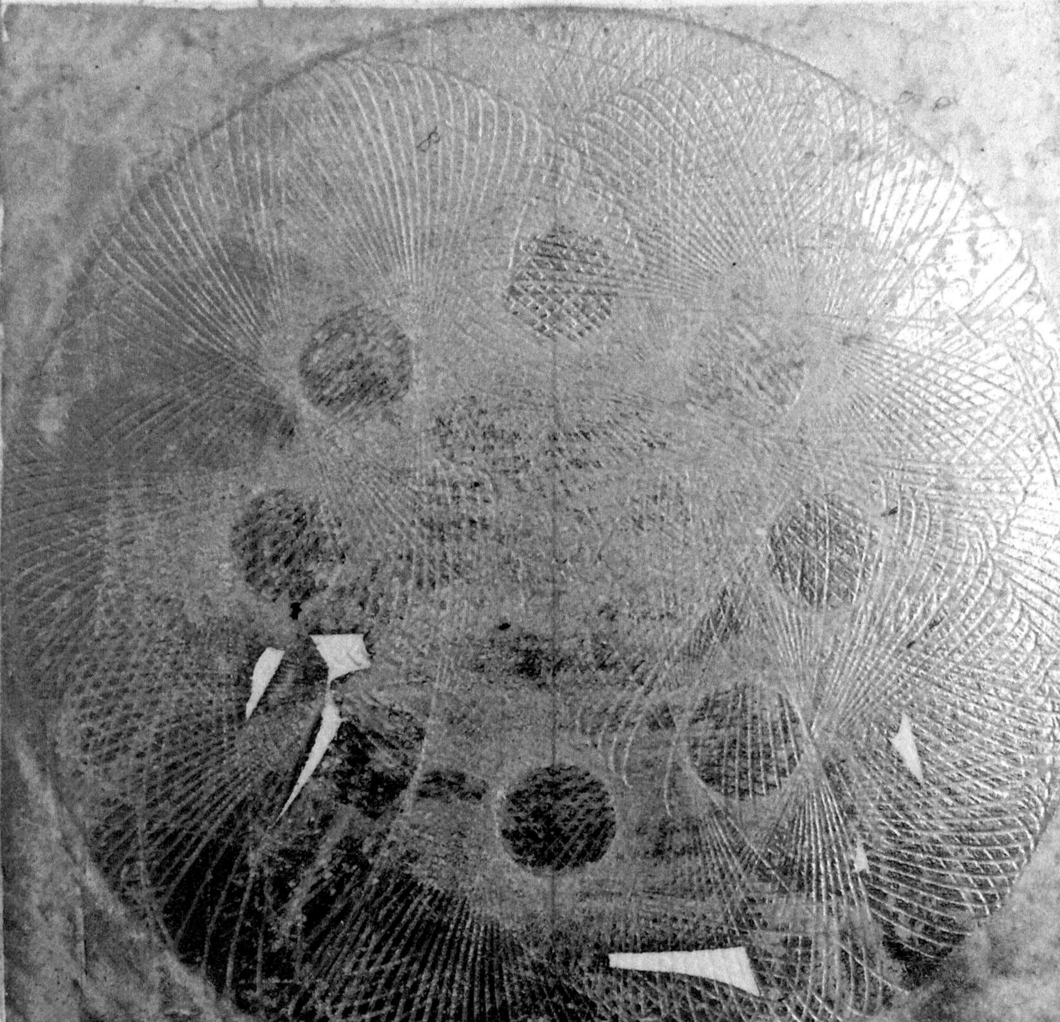

Figure 4.8 *Tangled Time*, *c.* 2015. Pencil, lacquer and silver leaf on paper, 25 × 35 cm. © Lesley Halliwell. *Tangled Time* began with a gilded surface. As I worked with the metal nib of a ballpoint pen and a burnishing tool, the surface was worn (away) to the point where it is uncertain whether the drawing is on top of the gilding or the gilding on top the drawing. The eternal return of marking, erasing and remarking creates a surface that can never be finished. While the outer–inner or above–below distinction offers us an organizational framework through which to discuss some of the complexities of surface, I am aware that such terms can suggest a series of strata. Surface is much more than a simple layer of coating; nor is it something that can be peeled off, as in an archaeological excavation. The depths of surface are complexified in relation to the position of the viewer, as a disjunction in the fourth dimension leaves a shadow in the surface.[1] And then there are also light and reflection, which destabilize the view, creating moments when the surface appears to hover or dissolve. The artworks discussed in this chapter operate in the realm of the expanded surface – as layered erasures, as layered time, as layered distances – that cannot be conflated into a determinate dimension.

Note

1 I thank the anonymous reviewer whose comments helped improve this chapter.

List of Figures

Chapter 5

Where Surface Meets Depth: Virtuality in Textile and Material Design

Elaine Igoe

The 'swatch' or sample format has been a common starting point throughout historical and contemporary practice in textile and material design. There are some basic conventions in the disciplinary tenet of textiles for the design of swatches and samples, which persist in design education and the commercial industry: 'trim it so there aren't any borders – make your design appear to go off the edge'; 'It mustn't ever look like a picture, but it should usually be rectangular whatever size you choose'; 'Make it look like it goes on and on and that you can buy as much of it as you want'; 'Mount it on card with a large border but only stick it down along one narrow edge so that it can move and be handled'; 'It may already include a repeating pattern through its construction but, if not, at least give an indication that the design repeats'; 'They will put it in repeat; it doesn't always matter if you can't wash it. It's about the idea. It'll be used as inspiration' …

Materials are both formed as, and cut into, rectangular formats for the purposes of understanding their varied properties through viewing and handling. Although swatches and samples vary in size depending on certain factors – such as the intended application, the techniques and processes used and commercial contexts they will be involved with – they mostly exist and perform as rectangular forms in trade shows, material libraries, studios, workshops, factories, shops and homes. In the chain of design, textiles and materials hold a foundational yet liminal position linked to their changing form as and within other designed objects, shifting from a sample or swatch to an applied surface, or material form and/or representation of an(other) designed object. Textile and material design practice has and is expanding into an ever-broadening realm, encompassing cloth, chemistry and computation; the universalism of the way we view, handle and understand the 'swatch' in relation to these changes in practice invite discussion. Atkinson et al. (2013) consider the 'tactile perceptions of digital textiles' in contexts where digital textile swatches are the communicative tool in a design scenario. Bruna Petreca's (2016) doctoral work recognizes the affective experience of handling and selecting textiles and materials and seeks to develop a framework for textile selection within the process of designing. These research projects develop ways to understand sensory perception through understanding materials and their properties as textiles.

This chapter deepens and balances this approach by developing a wider theoretical view of how virtual textiles and materials actually deliver affect. To do so, I begin by mattering, or giving importance to, the textile swatch or sample, uncovering its premise and critiquing its rectangular format as framing. Notions of framing are developed through the work of Anne Friedberg and her book *The Virtual Window* (2017), in which she uses the metaphor of the window or opening as a multiperspectival framing device. Explored in the context of a post-digital viewpoint – one that recognizes the prevalence of computation in daily life – the framing of textile design is here developed into concepts of virtual windows and manifestations of computation. I identify five contemporary studios whose practice navigates the blurring boundaries of the actual and the virtual and the immaterial and the material. Their work represents textiles and materials with agency across both the actual and the virtual, utilizing material and immaterial design practice in a co-emergent and critical relationship. They are: Molly Smisko's mixed reality experiences which critique our relationships with elemental materials; Lucy Hardcastle's design of material and immaterial surfaces; Emilie Carlsen's generative method of processing digital waste as textiles; Zeitguised's digitally rendered textiles which defy convention; and Angella Mackey's work which foregrounds the performance of textiles in the contemporary post-digital environment. In so doing, I seek out elements of a common framework to support and firmly bond these varying forms of practice as 'textilic'.[1] What forms a framework are the structural elements that persist, and this chapter focuses on two simultaneously: *form* – the rectangular framing device of the swatch and the screen – and *materiality* – the relationship between tangible samples of matter and representations of intangible matter.

Mattering samples

Rails upon rails, piles upon piles, textile designs are ready and waiting. Taking up their pre-destined role as one choice out of many. Their colours, their motifs, their textures, their fibres, their weights offer promise in plentiful supply. They are presented as 'samples' or 'swatches': the archetypal rectilinear format of cloth used in manufacturing and commerce. When designing and making a textile, we think and work through rectangular or square structures – the loom, the silkscreen, the computer screen, the digital drawing tablet, the sheet of paper, the squares of a grid. For the purpose of exhibition or to invite trade, we present them also in rectangular formats, sometimes draped, sometimes flat. Rectangles, both rigid and soft, surround us. They are doors, windows, books, screens and textiles. Even bio-materials, such as kombucha 'cloth', are bound by the size and shape of the vessel or tank in which they are grown or cultured. Textiles, as lengths, samples and swatches, even when freed from the devices which make them, are bounded. Each swatch or sample contains design. Each represents a design process. Each should be understood as a designed object and concept in itself. However, the role of these rectangles of design are to deliver an example of properties and capacities that could exist in the world, in whichever way you imagine them to. Some experience this is as the 'tyranny of the swatch' (Laughlin 2010), experiencing feelings of frustration caused by the limitations of bounded materials, which leave too much to the imagination, feeling they are not materially enough to be fully understood.

Fabric, frames and Friedberg

To begin her book *The Virtual Window*, Anne Friedberg quotes from Leon Battista Alberti's Renaissance treatise on painting and perspective, in which he uses the window as a metaphor. Alberti sees the window as an aperture, opening and closing, separating the spaces of here and there. The window is also a membrane, 'where surface meets depth' (Alberti cited in Friedberg 2006: 1). It is a frame, holding a view in place. Friedberg (ibid.) expands and challenges Alberti's window of single point perspective when she considers it as a screen; at once a surface and a frame – 'a reflective plane onto which an image is cast and frame that limits its view'. She sees this screen as a 'virtual window' that changes notions of materiality, space and time. Alberti's window is not a transparent window on the world but a 'windowed elsewhere' – a virtual space that exists on the virtual plane of representation (ibid.: 243). Friedberg stresses that the use of the word 'virtual' cannot be understood in its digital point of reference today but in its classical root in the word *virtus* (ibid.: 8), Latin for 'strength' or 'power'. In exploring the word 'virtual', Friedberg reminds us that that which is virtual possesses a power of acting without the aid of matter: '[a]n immaterial proxy for the material' (ibid.: 8). In this sense, virtual imagery has a 'second-order materiality' (ibid.: 11) and a liminal immateriality. Friedberg encourages freeing the term 'virtual' from its digital or 'information age' connotations, so that its 'material' connotations become more effective. Establishing this distinctive definition of the virtual helpfully illuminates a historical continuity between the virtuality produced through mirrors, paintings, camera obscura, photography and film, and how we largely understand the notion of the virtual today.

Friedberg defines how the term 'virtual' distinguishes between any representation that appears '"functionally or effectively but *not formally*" of the same materiality as what it represents' (ibid.: 11). Therefore, she describes the 'virtual window' as both a metaphoric window and an actual window with a virtual view; the metaphor functioning as the point of transference into the virtual (ibid.: 12).

Textiles framed as rectangles can be seen as the point of an 'ontological cut' (ibid.: 5). As Friedberg suggests, the cut occurs between the context of the material surface and the experience contained within its aperture, but also between the material surface and its virtuality. Tim Ingold (2007b: 1) – via Gibson's (1979) work, *The Ecological Approach to Visual Perception* – discusses the tripartite division of inhabited environments into mediums, substances and surfaces, and argues that materials are bound up in and between these, continually evolving in the 'flow of the currents of the lifeworld'. In describing the properties of a material in an environment, we are directly experiencing their *occurrence* within a lifeworld rather than the existence of an objective or perceived material with an intrinsic agency. Exploring and utilizing broader understandings of material as an occurrence, in combination with Friedberg's window metaphor through the schema of textile design bounded within the sample or swatch (or indeed screen), is useful. In doing so, we are drawing attention to the second-order materiality and liminal immateriality of conventional *and* emerging forms of textilic design, affording continuity between these practices.

Swatches as spiracles

A swatch is a portion of cloth, a piece of (or a sample of) fabric. The etymology of 'swatch' uncovers a shift of meaning, from the tally or tag attached to cloth to be dyed to the detached sample piece of cloth itself. The idea that the piece of fabric we handle as a swatch can always be considered as having a counterpart elsewhere – from which it is detached – sets up the established understanding of a textile swatch as a representative; both a promise and a possibility. The word 'swatch' and the term 'swathe' (connoting plentifulness) are both related to the noun 'swath' which at once points to a space covered by the single cut of a scythe ('a strip, lengthwise extent') as well as the agency of that space as a trace or vestige. This is poetically reminiscent of Simone Weil's (cited in Scarry 1998: 77) 'tear in a surface': a tear, brought about by the action of tearing, produces both a 'thing' and a space; 'Small tears in the surface of the world that pull us through to some vaster space'.

The connection between objects, words and their meanings offers us new ways of knowing new forms of objects in new times. If we reconnect with these etymological definitions, a rectangle of cloth can be understood as representative of the presence of the material and the absence of the immaterial, simultaneously promising plenty and remembering traces. A piece of cloth tacitly promises the rest of the cloth, in its design and making and its application and use. The rest of the cloth is diminished in its material form by the absence of the piece yet gains new meaning through this separation. They operate as a material–immaterial continuum. When understood in this way the shape of the swatch is an opening of possibility and memories. These thoughts also draw us to the notion of the *spiraculum eternitatis* ('a breathing hole into eternity'), developed in the sixteenth century by alchemist Gerhard Dorn (Fremont 2017: 2). Just like Weil's small tears, Dorn's spiracle ('breathing hole') is understood as a window-like space into other realms. It is a threshold to alternative knowledge which can thus be manifested through functions of *translation* and *art-making* within our tangible existence (von Franz cited in Fremont 2017: 2). Dorn's idea of the spiracle was an effort to 'translate the untranslatable' and so, as is the nature of translation, this threshold allows movement in both directions and also in concurrence. Translation can be understood as a function of designing; translation and designing both employ decision-making and expression for the purposes of communication (Baule and Caratti 2017: 15).

In conventional textiles, designers are translating a narrative or 'mood' through the selection of the combination of materials, forms, symbols, colour, context and function. This highlights the semantics of textiles (Andrew 2008), which develops from the use of repeating motifs, texture and colours. The patterning and repetition of symbols, motifs and motives coordinate with the window-framing effect of the rectangular format to again indicate a space of possibility, which is both inviting and disorientating.[2] The repetition of pattern is momentarily truncated by the methods of framing, thereby affording and enhancing the spiracular experience. Absence here connotes possibilities and the spectral presence of immanent counterparts. This alludes to the operation of 'textasis' (Igoe 2013). 'Textasis', the tension of 'textility'[3] (Mitchell 1997; Ingold 2010), recognizes etymological and metaphorical connections between text and textiles, between thinking, speaking, writing and making. Textasis suggests a movement between *stasis/enstasis* – unmoving, immobilized, subordinated, standing firmly within oneself – and *ex-stasis/ekstasis* – flow, excess, ecstasy, joy, insubordination, to be outside of oneself and the transgression of boundaries. In this sense, the rectangular swatch or sample can be understood as a spiracular framing device through which textasis plays out. The swatch/window, in the movement of textasis, becomes a frame for the act of designing textiles.

And so, the significance and role of edges, borders and surfaces as framing devices for textile designing, as the threshold between both the material and immaterial constituent parts, must be explored. The movement between material and immaterial through the swatch operates apparition-like: not always clearly visible, particularly once a fabric has been further cut, sewn, digitally rendered, stapled, glued and applied within the making and designing of other objects. It is thus important to explore what textiles can do before they are cut and sewn and in 'situ' (Lottersberger 2012: 46). In *The Material of Invention*, Ezio Manzini (1989: 183) prompts us to understand the primacy of a surface as well as 'the dynamic qualities that are concentrated in the surface'. He goes on to assert that the idea of static borders or edges of matter – in this context, the edges of a framed painting or a piece of cut cloth for example – is replaced by an idea of the surface as an interface, enabling an exchange of energy and information between the substances put into contact with it. He cites architect and designer Andrea Branzi (cited in Manzini 1989: 196):

> Unlike the surface of a painting, … a decorative surface implies infinite borders, and contains in each of its smallest parts the sum total of information in the entire system, since it contains the individual sign that is then repeated *ad infinitum*.

Manzini (1989: 192) also recognizes that repeated signs can create a rhythm. Repetition and rhythm indeed remain some of the characteristic features of decorative surfaces. Returning to the various framing devices through which the action of textile designs operates, repetitive and rhythmic action – spinning, weaving, stitching, knitting, printing, pasting, coding – translate creative processes into rhythmic texture, motifs, and compositions that 'go off the edge'. These repetitious aspects serve as signifiers of the creative action of the maker as well as of the material–immaterial tension within textiles. Our visual and sensual perception implicitly manifests the apparition of the immanent counterpart. Just a small cut piece of textile is enough to do this.

Post-digital textilic works

We live in a post-digital context where notions of the digital and the analogue intertwine in an increasingly sophisticated way. The post-digital defines a position in which the novelty of the digital has been overcome and its value is becoming fully integrated into our lives (Openshaw 2015: 5). Florian Cramer (2015: 17–18) reminds us that the digital need not be computational and the analogue can indeed perform computationally. Post-digital textile design utilizes both digital and analogue techniques and practices to create outcomes such as bio-design, virtual reality works, smart interfaces and materials, as well as print, embroideries, weaves and knit. The shared reference points and language of the digital and the textile have been stated before, pioneered by Sadie Plant in *Zeros + Ones: Digital Women and the New Technoculture* (1997). Textiles, in their typical form, can be understood as digital at the time of their production: numbers of counted warp strands, a predetermined number of stitches. After all, the 'digital' denotes something divided into discrete units; with the noun 'digit' denoting a finger or toe. A digital system can be basic or highly complex, but it is systematic and traceable. What makes textiles analogue is our relationship with them, the way we interact with them, wear them out, imbue ourselves into them. Yet once out of use they can be unravelled; each strand or row once again can be counted. Cramer (2015: 18) points out that '[t]he structure of an analogue signal is determined entirely by its correspondence (analogy) with the original physical phenomenon which it mimics'. As is understood, textiles and cloth were archaically created to mimic natural surfaces such as hair, fur and skin, not only for their function but for their sensorial qualities (Eadie and Ghosh 2011). And so textiles, which at their point of production can be considered digital, ape the analogue qualities of the body. This analogy then evolves ambiently to become a sort of meta-analogue surface. The affect of textiles is then derived through their virtuality as defined by Friedberg. Post-digital textiles and materials derive agency from notions of the surface as a designed object throughout cultural history, as well as the significance of their virtuality and existence as a networked representation.

We experience a significant proportion of post-digital culture via computational screens. These screens are reformulating our view and experience of the world and our ability to act on it and within it (Openshaw 2015: 9). It is indeed already possible to interact with computation through a range of materials, from Jacquard by Google – a woven fabric interface already available commercially as garments – to Caroline Yan Zheng's (2019) research into soft robotics, where the tactility of the materials is the primary route to engagement with computation. However, the screens we still mostly use are usually made of hard, flat glass. Our tactile relationship with them is limited, but the surface absorbs and transforms us. When using computer screens, we are used to seeing multiple, framed and layered images. We can choose how to work with them, inviting a visual hierarchy or seeing all at once. A screen as a surface is fractured by and given depth by that which takes place within it. This represents a 'remade visual vernacular' which 'requires new descriptors for its fractured, multiple, simultaneous, time shiftable sense of space and time' (Friedberg 2006: 3). In post-digital design practice, screens and materials can be said to co-emerge as integral sites of design processes, leading us to question where the boundaries between screens and materials are – if any exist.

This notion of co-emergence is fundamental to David M. Berry's (2015: 44, original emphasis) concept of the post-digital 'constellation': 'The postdigital, as an aesthetic, gestures towards a relation produced by digital *surfaces* in a bewildering number of different places and contexts.' Berry refers to Bruno Latour's (cited in Berry 2015: 45–6) critique of the digital, in which he denounces the abstraction of the digital and draws attention to its basis in the material and socio-technical world: hardware, silicon and binary code. In developing his concept of the post-digital constellation, Berry condemns Latour's understanding of the digital as a set of observable practices in complex data. For Berry (2015: 47), this is insufficient in representing the contemporary omnipresence of computation and the relationship of the digital and the analogue. It only interrogates a fraction of the 'digital iceberg' which is created through layers of abstraction, from material manifestations of computation through to digital traces. In the post-digital constellation, the digital must be understood as a historiographical spectrum layered and entangled with the social and material analogue in highly complex ways.

Post-digital textiles exist along a continuum of forms of practice. They often exemplify textasis in their oscillation between the digital and the material, performing the spiracular function in both their process and outcomes. In qualification of this, five examples of post-digital textilic design practice are presented below. In these examples, intangible work is anchored in the material world through the design and making process or at the point of interaction via mixed reality. Material forms are augmented, altered, tested, mimicked or made 'virtual' through computation.

Molly Smisko of Smisko Ackerman is a digital artist based in New York and was trained in textile design at Chelsea School of Art and Design in London. She has developed work in mixed reality (XR) like *Assimilation* (2016). The person experiencing this example of design enacts the properties of the spiraculum. Their body is the interstice or interface between the tangible and the intangible. In *Assimilation* we see a stark surreal landscape in VR, while the tangible environment is rather bland. In Figure 5.1 we see a wall-mounted fan, a bowl of sand and a flimsy white curtain. The main image is the virtually

Figure 5.1 *Assimilation* (Screenshot), *c.* 2016. © Molly Smisko.

inhabited environment. The experiencer sees and hears the familiar effects of the elements: wind blowing sand, water rippling and bubbling. In the tangible environment we feel sand, we feel water and the movement of air across our skin. The properties of these materials, both tangible and intangible, augment each other and force us to question them too. The gap between the digital and analogue worlds are exposed affording a post-digital critique of these elemental materials. Smisko also creates digitally rendered, animated textiles. In *Membrane* (Smisko Ackerman, 2018), a pulsating, ruched body of digital fabric is seemingly internally lit and expands and contracts alarmingly, moving towards us of its own volition within the frame of the screen. We then see folded materials encapsulated within a liquid form, as well as shiny visceral surfaces being forcibly stretched by something (a hand maybe) off-screen and out of our frame of vision.

Lucy Hardcastle is a London-based digital artist with a background in both textile design and information experience design at the Royal College of Art, London. Hardcastle's practice blurs the tangible with the intangible. She uses sketching, glass blowing techniques, 3D printing, flocking, hand dyeing fabrics, 3D rendering, digital animation, photography and sound effortlessly to create 'real and imagined touch, visual illusions and sensual aesthetics' (Hardcastle 2019). Liza Mandelup's profile on Hardcastle (2016) shows her at work, shifting through the various 'frames' of her practice. Her rectangles of cloth, selected for texture, dyed to a specific shade, expand and are extracted into their material properties through her practice, much in the sense of an exploded technical drawing. These material properties tell 'condensed stories' (Ingold 2007b: 14) and move into new realms as they are rendered digitally, taking on new forms which do not behave in ways that we have come to expect in the flows of mediums, substances and surfaces in our tangible environment. Rather they perform in new ways, dictated by different forces – algorithm, code, hardware, screens, eyes. In *Archetypes* (2019), fibrous textures are seemingly growing on ice or some other crystalline structure, then tumour-like growths appear and become part of the thing which captivates us, wrapped in tendrils. These glistening and furry surfaces seem to crackle, melt and yet grow before our very eyes (Figure 5.2).

In a blog post for the exhibition series *Rethinking Matter*, curator Pamela Grombacher (2017) uses the work of textile-designer Emilie Carlsen as a discussion point for digital materiality in reference to Paul M. Leonardi's (2010) article 'Digital Materiality? How Artifacts without Matter, Matter':

> Her textiles are not the simple inscription of data, but are instead the product of a mutually constitutive exchange between data and fabric. She materializes data as textiles, and then digitizes these textiles to create new data, blending digital and analogue processes in a back-and-forth that blurs the material and the immaterial realms.
>
> (Grombacher 2017)

Figure 5.2 *Archetypes* (screenshot), *c.* 2019. Lucy Hardcastle Studio for Nordic Health Lab. © Lucy Hardcastle.

Grombacher suggests that Carlsen's textiles defy Leonardi's depiction of data as immaterial, unless it is manifested materially. Her practice leaping in and out of the spiracle, Grombacher describes how Carlsen begins in the digital realm with what she calls 'digital waste', working this into designs which she then prints onto different materials to create varying textural and colour effects. She often then rescans these printed materials to produce new permutations of the digital imagery (Figure 5.3).

Zeitguised are a multi-disciplinary, Berlin-based digital studio, who specialize in the creation of 'exquisite realities' which carry 'hallucinative narratives of shape, color and behaviour relations' (Zeitguised/F°am 2018). Their archive of work includes a number of projects where digitally rendered textiles are the focus. *Emancipath* (2018), a collaboration with Danish interior-fabric-manufacturer Kvadrat, is an interesting exemplar. Kvadrat are known for their high-quality canvas textiles and in this film Zeitguised introduces new qualities to these textiles through the means of '3Dcgi craft'. Canvas becomes flexible, fluid, stretchy and it bubbles. Organic rippling shapes with 'fabric' surfaces detach from angular foundations: an emancipation. They play with our understanding of how textiles *are.* They describe how they have knowingly moved the tiled textures across the geometric shapes to expose the work as a digital construct. This act of post-digitality in turn forces us to question known rules for dealing with tangible textiles: fabric grains running – literally moving and flowing – in opposing directions, the complexity of cutting and piecing of cloth to fit around 3D and moving shapes, weave structures that should not stretch, but do. In the work of Zeitguised, these are arbitrary restrictions. In this piece, the edges

Figure 5.3 [*left*] *Digital Realism*, *c.* 2015 (Emilie Carlsen); [*right*] *Melt & Joint*, *c.* 2018. FLOCK (Emilie Carlsen and Siff Pristed), Japan. Photograph: Lasse Kusk.

are what instantiates these questions. The edges/selvedges are moved to the centre of the frame in an act of folding (Deleuze 1988). The capabilities of this immaterial cloth, though recognizable through a weave structure, is unknowable, unpredictable and unreal (Figure 5.4).

Angella Mackey is experienced in the design of physical–digital hybrid fabrics. Zeitguised goes as far to call these 'phygital'. Mackey undertakes 'design explorations' for dynamic fabrics that have the abilities of a computer screen. She asks, '[w]hat would it be like to wear fabrics like this?' In her 2019 work with Phem, she wants to show us, asking us to choose a video we want to wear. Mackey uses live green-screen technology to facilitate this dynamism. In the work of Phem, phygital textile designs come about through sym-poiesis. Tangible fabrics are designed or chosen using specific colours or shades, and the semiotic propensity of these colours is expanded in the digital realm. Mackey designs 'digital shimmers' which are the animated intangible layers that will be incorporated onto real cloth and into our sensory perception through the rectangle of a smartphone screen. In an archetypal post-digital approach, Mackey produces all this through off-the-shelf apps, purposely exposing the digitality and encouraging further processing and digital granularity by filming screens showing films.

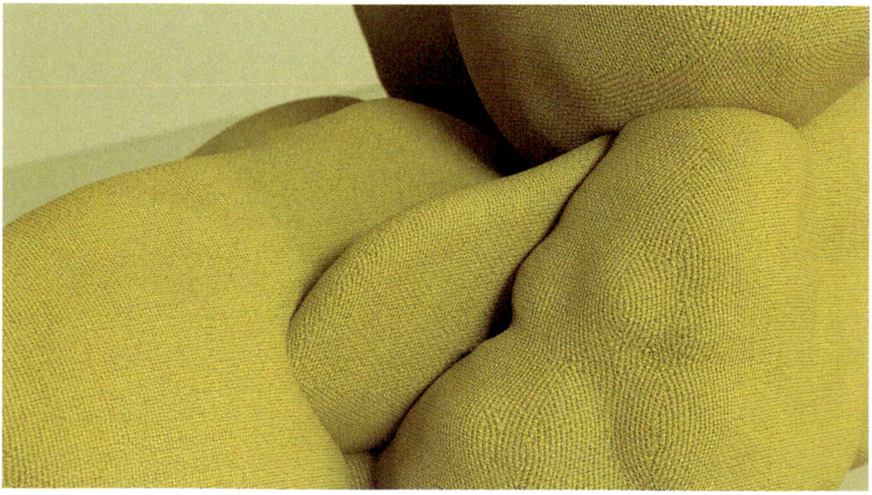

Figure 5.4 Emancipath (Screenshot), *c.* 2018. Zeitguised/F°am in collaboration with Kvadrat. © Zeitguised.

These textile practitioners are playing in and through the spiracle, layering up performances of textilic design. This layering, coupled with concepts of frames and windows, creates a *mise en abyme* (Friedberg 2006: 14), functioning as nested apertures fading from view. This nesting does not simply suggest a metaphoric scaling, but an extension, multidimensionally, into virtuality.

> Matter, in our view, is an aggregate of 'images'. And by 'image' we mean a certain existence which is more than that which the idealist calls representation, but less than that which the realist calls a thing – an existence placed halfway between the 'thing' and the 'representation'.
>
> (Bergson [1908] 1998: 9)

Immaterial textiles

After considering the work of these practitioners, it is easy to accept that the digital can have material qualities. Their ability to deliver affect is evidenced above. Virtual textiles and materials are wholly affected by the flows between medium, substance and surface: the quality of the hardware being used, the speed of the data connection, the sensory preferences of the experiencer. I suggest here that textilic practice, often manifested conveniently within frame-like geometries, can offer a spiracular experience – an opening which affords types of experience belied by their tangible forms.

In the work of Smisko Ackerman, Lucy Hardcastle and Zeitguised, the use of sound is particularly affecting. The sound we associate with tangible materials is most often instigated by our interaction with them. Soft crispness, a hollow wobble, a squeak, a subtle or zipping sound caused by friction against skin or a nail … silence even. I assume that, as we cannot actually touch works of immaterial textiles, sound is incorporated as part of the design process. In the examples given, it appears that these sounds are also created through computation, containing aural motifs and repetition, and sometimes with no obvious beginning or end. There is a knowing and experimental choreography between the sound and the immaterial materials. Sound here is a contributor to the digital materiality we are experiencing. Contemporary electronic music is often created entirely digitally, sometimes imitating or directly using recorded sounds made by people and instruments in other spaces and at other times. New Materialist scholars Heidi Fast, Taru Leppänen and Milla Tiainen (2018) attend to the physicality of sound through vibration: vibrations engage materials, bodies and spaces in

> [a] mutually affective field of resonance, which surpasses, yet also reshapes, the subjective experiential register. … the vibrations of sound waves attest to, animate and enhance the liveliness and agential capacities of matter. … vibrations focus our attention to the interconnectedness and co-occurring of human and more-than-human materialities, bodies and entities.

Vibrations are transversal and disperse into other sensory registers (ibid.). And so sound, or the vibrations caused by sound, may be a more direct access point to the materiality of digitally rendered textiles than the visual. Vibration is of course recognized as a type of 'distance touch' or a 'structured pattern of pressure changes' (Katz [1925] 1989 cited in Ratcliffe 2018). These works bring the experiencer's body in direct contact with digital materiality, thereby emphasizing the quality of 'stuff' within complex assemblages (Deleuze and Guattari 1987; Delanda 2016) or constellations (Deleuze and Guattari 1987; Berry 2015).

I return to the passage at the beginning of this chapter, purporting to the 'conventions of textile design'. With our developed understanding of immaterial textiles, these conventions take on new meaning and emphasize the expanded practice of textile design; that is, textilic design in the post-digital. Rectangles of fabricated fabric, textilic works, become tears in our collective consciousness, framed so as to become windows into other realms. Swatches and screens both offer a virtual materiality where presence implies absence and vice versa. Post-digital textilic works occur as condensed stories delivering sensory affect. Through rectangles, we experience embodied, enacted information. This information exists as a counterpart, within *and* beyond the material itself. The consideration of the presence of the absent counterpart can be understood as a substantive element of the act of designing textiles. Whether the textile is created through rectangular silkscreens or computer screens, they can offer information, experience, affect and possibility across the borders of their material and immaterial existence.

Notes

1 'Textilic' (Ingold 2010: 91) defines a material-led making process, which through Renaissance treatises became marginalized by the architectonic and hylomorphic models.

2 Jane Graves (2002: 47) mentions the Freudian unconscious delight and anxiety-inducing properties of pattern and repetition in textiles, creating a feeling of being 'on the edge of a nightmare', while using a dislocation of scale to 'manage' these qualities within the rectangular framing of a piece of cloth.

3 'Textility', in the work of both Victoria Mitchell and Tim Ingold, defines a contiguity between textuality and tactility. Textasis explores the activity manifested from that contiguous positioning.

List of Figures

Chapter 6

Growing Surface between Textiles and Electrochemistry

Joanne Horton

As a mixed media textile designer, I work up surfaces layer by layer responding to the physical experience of making and the personal relationship I have found between materials, surface and structure. I also see myself as a lay electrometallurgist, learning electrochemistry and applying that knowledge to textile processes.[1] Through an 'investigative artisanal practice' with a strong affinity with science, I devised a special type of metallized embellishment: one that is *grown*, not sewn. In this chapter, I will first discuss how my interest in surface has been inspired by the nineteenth-century British industrial technology, electrometallurgy: a process that blurs the line between alchemy and science, and between art and science. Next, innovative links will be made between supple textile and resistant metal, through electroforming with copper as 'the art of growing parts' (Sole 1994: 29), unsettling the conventional living–non-living distinction. From this, I will discuss the playful approach to my practice – combining old and new technology, retrofitting machinery and testing out ideas in unfamiliar fields – that has led to the production of sumptuous and unexpected metallic textile surfaces.

My practice is influenced by history and driven by materials. I look back to look forward, trusting that past technical know-how bridges experience and innovation. My initial interest in electrochemical metal surface arose from a fixation with alchemy, the experimental probing of Queen Elizabeth I's alchemical-occultist advisors and their quest to make base materials 'precious'. While inspired by esoteric scientific process, I uncovered a more pragmatic and striking account from Harriet Martineau's 1851 account of watching electroplaters as they worked their 'magic troughs' to create many different types of decorative objects: '… [the magician] takes hold of one of the wires and unites it with the wire on which the goods are hung. Then in an instant they become overspread with silver' (Martineau cited in Patterson 2012: 65). The transmutation of base metals by coatings of silver and gold became so universally fascinating that electrometallurgy is still referred to as 'almost magical' (Jones 2012) or even as a 'dark art'.

In my experiments with electrometallurgy, highest-quality polyester tulle was selected as a base on which to 'grow' copper by electroforming. The liquid-looking metal copies the structure of net as the result of its unique chemical-metal make-up. My preoccupation with electroforming was rooted in its lack of familiarity as a textile working process – except for stiffening and 'bronzing' baby shoes and lace, being made waterproofed and then metallized to make jewellery – and in the unique transition from chemical-metal particle substance via electrical attraction to tangible metal surface. Electroforming and electroplating are methods of covering objects with layers of metal, using electric current connections, to move metal particles through a water-chemical-metal salt solution (electrolyte) held in a tank to deposit on the selected object. In electroforming, the deposition of metal forms is made particle by particle, layer by layer, in metal lattice shapes, growing thick and strong enough to create a self-supporting object. In electroplating, by contrast, a thin layer of precious metal is applied to the base shape.

Initially I wanted to understand the electroforming process and its potential to build surfaces in a 'magical' way: the scene of removing a finished, solid, red copper electroform from its bright-blue copper sulphate chemical solution seemed to bridge alchemy and modern science. To apply this process to textile surface, however, I first had the practical task of making non-metallic textile substrates electrically conductive. Figure 6.1 is the result of drawing on 100 per cent polyester fibre using a pen containing silver powder in medium. The pen is electrically conductive to draw metal through the electrolyte to the textile surface. This is then incrementally grown by a specialist acid formulation of copper, as a bright, shiny, red wall built in layers, that requires no finishing such as sanding or filing. The resulting surface can be made thin, light and flexible, or thick and heavy, as desired.

Figure 6.1 Electroformed copper on polyester tulle (detail), *c.* 2016. © Joanne Horton.

Over two decades ago, Matilda McQuaid (1998: 19) speculated that dissolvable fabric and layering processes may alter our ideas about embroidery. This view prompted me to explore the creative potential of technology and material science in the production of embellished textiles, particularly in combining cloth with metal. If successful, it could offer alternatives to metal thread embroidery or goldwork, a new type of layered surface processing situated between print, embellishment and chemistry. In the past twenty years, a quartet of textile designers – Frances Geesin (1995), Kinor Jiang (2009), Tine De Ruysser (2009), Sara Keith (2010) – established a new field of decorative electro-deposition on textiles. They interrogate ways of building up surface by coating with metal, hand-painting and other techniques to isolate areas against metal deposition using electroforming, electroplating and electroless plating.[2] The creative interplay between soft and hard surface features, in Tine De Ruysser's work for example, involves applying metal to strategically processed areas and folding the remaining pliable areas of fabric, to form three-dimensional geometric shapes. I was inspired to explore the play of softness/rigidity, roughness/smoothness and lightness/heaviness through electroforming in such a way that the textile substrate remained soft and draping even when metal was grown onto it.

Although copper is known as a base metal, Aldersey-Williams (2011) comments on its special status in relation to gold, as the only other truly coloured metal. This held my interest, together with the novel reflectance and bright colouration of acid copper electroforms, thanks to the presence of organic surfactants in the electrolyte solution that combine with a very pure form of copper during processing. Furthermore, in textiles, copper offers a distinctive warm colouration and high gloss that can contrast beautifully with softer textile materials (Braddock and O'Mahoney 2005). I experimented with surface quality by growing jewel-like sections that are permanently attached to high-quality polyester organza (Figure 6.2). I was intent on encouraging copper to embed into the structure of sheer lightweight fabric. My selection of substrate was informed by the handle and drape found in high-quality tulle, bonded web taffeta, organza, imitation silks and chiffons – from relatively stiff to fluid, and from matt to shiny. I was mindful of the unique capacity of an electroform to replicate and exaggerate its host surface. This is particularly evident in Figure 6.1, where rounded low relief walls of metal have grown, replicating the hexagonal format of tulle and exaggerating the fabric structure.

I filmed my designs forming as they grew inside the electroforming tank and the recording reiterated the importance of material and its relationship to surface. Watching back the intensely blue crystals and metal particles transforming into red copper layers through the (al)chemical reaction of electro-deposition led me back to the notion of growth. In *Making and Growing* (Hallam and Ingold 2014), various insights are given into what it means to make and work with natural resources or bio-engineered materials. The paradox of 'unnaturally grown natural surfaces' is fascinating, given prevalent references to electroforming as a 'dark

Figure 6.2 Electroformed copper on polyester organza (detail), *c.* 2016. © Joanne Horton.

art'. The copper sulphate crystals I used grow in a tree-like manner, forming a structure which underpins the electroforming metal as it solidifies. Although non-living, the crystals grow 'to flourish, to increase gradually in size by natural development' (ibid.: 1).

My craft-science-technology undertaking moved through a bricolage approach, by customizing conventional tools from the electroplating and electro-finishing industries, and by combining textile technology with cutting-edge equipment from electronics and manufacturing. Through trial and error, retrofitting and redesigning, I tested, tinkered and adapted. Eventually I processed my samples in my bespoke electroforming set-up in the laboratory at De Montfort University. However, I started with a 'home' electroforming kit: a 'Heath Robinson' photo-etching tank adapted for electroforming (Figure 6.3, *right*). This simple tool allowed me to test out forming layers of copper on small amounts of cloth that had been prepared with hand-painted conductive ink to attract metal. I then went on to make a transparent polypropylene tank (Figure 6.3, *left*) to act as a subsidiary processing tank, holding just ten litres of costly electrolyte solution. I also fitted my main tanks (50 litres) with specialist frames known as jigs – commonly used to position three-dimensional jewellery items – to process textiles. Loose sections of cloth can obscure the conductive ink design on the fabric and affect the quality of the final electroformed design. By converting jigs into stenters (open supportive frames) on to which to peg

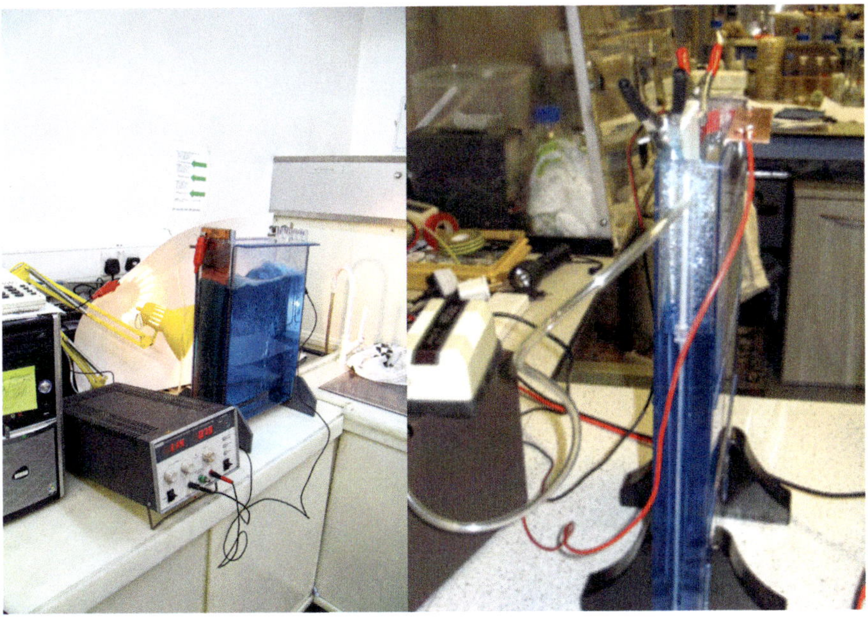

Figure 6.3 [*left*] Bespoke viewing tank, *c.* 2015; [*right*] Home electroforming kit, *c.* 2014. © Joanne Horton.

swatches, I was able to keep the textile tensioned in the tank. I also sourced conductive circuit pens used to teach children the principles of electronics and turned them into line drawing devices controlled by computer. This enabled me to create consistent, measurable designs in conductive ink on tensioned fabric prior to immersion in the electroforming tank.

In the course of my investigation, two particular challenges presented themselves. Firstly, to be able to use the right amount of current to produce a bright, shiny electroform, instead of a burnt one, it was necessary to calculate the surface area with complex, thin and flowing lines. This challenge was addressed by using a draughtsman's digital mapping pen to trace the exact surface area with designs prior to immersion in the tank. Secondly, the porous nature of unsealed fabric encouraged growth on the front and back of the selected surface, doubling the surface area during processing. This meant that the amount of power and time needed to create an effective electroform needed to be re-optimized. Rather than solely relying on calculations, I worked from my experience, removing swatches from the tank at appropriate intervals to adjust electric current.

The repeated process of testing and modifying resulted in a series of sheer net-fabric swatches with delicate electroforms in gentle curves, catching light in different directions. In these swatches, copper electroforms are guided by a

Figure 6.4 Filigree designs on polyester sheers, *c.* 2016. Electroformed acid copper. © Joanne Horton.

track of silver that activates the metal to flow across the surface, to eventually solidify, impregnating and copying the textile structure underneath. I also explored how different densities of metal lines can be processed to achieve a bright or semi-bright appearance (Figure 6.4) and found that hammering the metal post-treatment changes the way the surface captures light.

My experiments with material, structure and process have resulted in a surface with intricate metal detail that is delicate yet strong. The surface is flexible, capturing light and movement in multiplying folds. In the future, I would like to process electroformed garment surfaces in a tank. I have already tested this in a small section but aim to achieve far more sophisticated results. There are also opportunities to combine conventional embroidery on electroformed textile: used as the base, the strength of the electroform surface could function almost like the boning of a corset, in single or multiple layers. Moreover, as electroformed surfaces can be reintroduced to the tank to grow further, multiple layers could be joined to form a metal-textile sandwich or patchwork. I am fascinated by metallic surface, its patina and texture, and selecting a specialist form of metal to work with. As a designer from the Black Country, the heritage of metalwork and innovation in the area is inescapable. The metal-textile continues to grow, as does my empathy with material, structure and surface.

Notes

1 The term electrometallurgy was coined by Alfred Smee in 1840 to describe the coating and growing of self-supporting objects with metal by galvanic current using chemical solutions in a tank.

2 Also known as autocatalytic plating or conversion coating, electroless plating involves a catalytic reduction of ions to plate without any electrical energy dispersal. This is a purely chemical process with no machines or electrical power.

List of Figures

Chapter 7
David Pye's Fluting Engine

Benedict Carpenter van Barthold

Introduction

This chapter considers the fluted surface of a Brazilian rosewood dish (Figure 7.1) carved in 1980 by the British craftsperson David Pye (1914–93). The surface of this dish is inscribed by a unique hand-powered milling machine of Pye's invention, called the 'fluting engine'. This surface is an index of the operation of the engine, a system of making in which Pye (1986: 46) forms a human component, iteratively judging the spacing of the flutes by 'eye and hand' while the dish is being carved. The dish is part of a larger body of work made by the same process over a period of four decades. When this surface is considered alongside the evolution of the engine itself and the parallel evolution of gross form within Pye's work as a whole, the dish can be seen as a constituent of

Figure 7.1 David Pye, *Dish*, c. 1980. 420 mm long, Brazilian rosewood. © Crafts Council.

a larger and evolving system of production. This chapter therefore considers both the individual and the durational systems of production that are inscribed on the surface of the dish. The individual system is synchronic: it occurs at one moment in time and corresponds to the production of a single dish. The system is also diachronic: it is perpetuated over a long period of production. In this way, Pye's work forms a continuous project, composed of discrete instances that are distributed over time. The idea of the artist's body of work as a 'distributed object' is elaborated by the anthropologist Alfred Gell (1998: 232): 'It constitutes, as it were, an independent chunk of space-time, which can be accessed via each work individually, each standing, indexically, for all of them and the historical-biographical context of their production.' Following Gell, this chapter considers the surface of the rosewood dish as an index of a specific event of making and also of the extended system of creative evolution. These are contextualized with reference to Pye's theoretical writing and contemporary responses to his work. The chapter asks what knowledge we might draw from the example of Pye's practice, and how this knowledge might guide our use of new technology.

Pye's practice is not reliant on computer technology, but his work and ideas were developed at the new technological horizon of computer-numerical-controlled (CNC) manufacture. Since 2011, a number of makers have revisited Pye's work, either by attempting to remake his milling machine or by reimagining his work using CNC milling machines (Wumkes 2017–18; Grimshaw 2017; Kettle et al. 2014; Leonard 2011). This strange homage is fascinating, but the example of Pye's practice has not previously been considered alongside a fuller understanding of his theoretical writing.[1] This chapter presents a synthetic reading of Pye's theoretical work alongside his practice, to give fresh insight into the nature of practice mediated by machines, the improvisatory role of the human operator, the role of the hand in making, and the relationship between tools and human creativity. Pye's work is an object lesson and orientation point for makers who seek to extend their embodied knowledge through new digital technologies. By developing understanding of Pye's improvisation as a maker though his unique system of mechanized production, this chapter will inform research into contemporary incorporations of human improvisation and machine production, and our own horizon of machine learning. It is informed by my own experience of making objects, and the insights that I bring through imagining myself as the human component in Pye's system.

'Craft' is a contested term.[2] A common-sense definition is 'skill in making things by hand' (Frayling 2011: 9). This identity is perceived as increasingly valuable when it is challenged by cheaper and – it is assumed – inferior alternatives (ibid.). Craft historiography charts a succession of fraught pairings, such as with industry (Adamson 2013) or fine art (Adamson 2007). This narrative of constantly reimagined skill can be found in the craft historian Tanya Harrod's recent introductory overview, *Craft*. In this she writes that 'the realm of craft

[is] making ... something by hand' (Harrod 2018: 12), whether using traditional skills, such as glass-blowing, or the 'new kind of mastery of the hand in the iterative processes of programming' and rapid prototyping (ibid.: 16; see also DeNichola 2016; Sennett 2008: 24–7; McCullough 1996). But the challenge posed by computer technology to the crafts – and to the hand – is obvious. Either the screen replaces touch with sight, or, as touchscreen, it presents a surface on which the finger acts as an index of sight, a kind of touch without palpation, a dart of surrogate vision (Ingold 2017: 101–2). The new relation of maker, tool and material is governed more by eye than hand. In the words of the design academic David Grimshaw (2017: 3736), this is a diminution of the 'haptic and tacit knowledge ... that are the traditions of craft and material making'. At the same time, of course, craft practice has evolved to embrace rapid prototyping and digital design (Johnston 2015; Openshaw 2015). In this respect, the example of Pye's work and its modern reimagining is both test bed and guide: it has a contemporary relevance and instructive power well beyond what Glenn Adamson (2007: 71) described as craft's historical 'ghetto of technique'.

There is a persistent and popular notion, tinged with an incipient romanticism, that craft as handwork is an instructive retreat from the alienation and damage of modernity. This is evident from John Ruskin's *The Stones of Venice* ([1851–3] 2009: 151–230) to Robert Pirsig's *Zen and the Art of Motorcycle Maintenance* (1974). Over the last decade there has been an eruption of this belief in popular literature and in the first-person accounts of craftspeople (see Davidson and Tahsin 2018; Korn [2013] 2017; Crawford 2010). Because craftspeople have palpable ability in navigating the affordances (Gibson [1979] 1986: 127–43) of the world of stuff, craft has also been treated as an object of instruction by academic authors, as paradigmatic of aptitude in negotiating the world's opportunities and ambiguities. In Richard Sennett's book *The Craftsman* (2008), the socialist sociologist explores the idea that the hand – or *an idea of the hand* – might play a vital role as a paradigm for a self-critical merging of subject and world, fit for application in a range of contexts, from town planning to coding, architecture and politics. A more interesting and disenchanted example is provided by the anthropologist Tim Ingold in *Making* (2013). Near the start of his argument, he considers two different modes of perception: the optical and the haptic. The optical mode sees the end before it is reached, like an object on the horizon that is approached. This mode of perception structures human creativity as the 'novelty of determinate ends conceived in advance'. This is contrasted with a haptic mode, the 'improvisatory creativity of labour' (ibid.: 20). Ingold argues that our epistemology is dominated by a project-orientated idea of making. In this view, the maker begins with a mental image. Nature provides a supply of material. The project is complete once the material has assumed the image conceived in the maker's mind. This 'hylomorphism' – from the Greek *hyle* (matter) and *morphe* (form) – posits a hierarchy of form over matter. Ingold's intention is to explore a radically different way to understand making, one more attuned to a haptic mode of perception, to 'think of making, instead, as a process of growth' (ibid.: 21). Haptic engagement differs from project-orientated thinking in two crucial ways. The first is that the maker is continuous with the world. There is no way that an idea can be formed in advance of material. The second is that materials are not passive, but are in a process of continual movement, of growth and flux. A tree, for instance, can be seen to grow, and wood to age. But wood as a material for art is taken into culture's fold only through the agency of a sculptor: project-orientated thinking values the sculpture but not the tree. It is against this view, hegemonic in Western humanism, that Ingold proposes his idea of making as growth.

Ingold's work on making is informed by his fieldwork, secondary anthropological literature and various practical workshops. He is not a dogmatic author, but he can be abstract, and his brief attempt at, for instance, basket-weaving (ibid.: 22–4), instructive though it may be for his theoretical project, is of a rather different status from a lifetime's practice. If we want to understand how (and whether) his ideas are instructive in fact, we must find other examples. There are a few authors who write from the privileged position of being really expert at the things they make.[3] One such is the British craftsperson David Pye. Ingold mentions Pye in *Making*, but in relation to Pye's writing and not the objects of his craft (ibid.: 29, 62).

David Pye, in theory

Pye was a furniture designer and teacher who worked at the Royal College of Art (1949–74). He is best known today for his influential book *The Nature and Art of Workmanship* (1968). This book has dominated understanding of Pye's legacy. In it, Pye divides making into two distinct phases, placing ideation before realization, and following an optical mode of perception. But a fuller reading of Pye's legacy needs also to include his now less-regarded book *The Nature of Design* (1964).[4] It is the task of this section to read these two works together, to reach a better understanding of Pye's model of making. Pye is also known for a series of accomplished bowls, carved from wood by means of the fluting engine, invented sometime between 1949 and 1950. This is a complex tool by which a hooked gouge may be directed by a lever to cut the distinctive flutes that characterize his bowls. Many of Pye's bowls are now in major collections, including the V&A Museum (London) and the Museum of Art and Design (New York), as well as specialist collections such as the Centre for Art in Wood (Philadelphia). The Brazilian rosewood dish that is the focus of this chapter is in the collection of the Crafts Council in London (W43; see Crafts Council n.d.; Crafts Council 1986: 63; Fig. 7.1).

Pye is one of the few post-war British craft theorists of note. Definitions of apparently simple concepts such as 'skill' or 'craft' are hard to agree.

Pye recognized that these terms are freighted with meaning to the point of uselessness. Consequently, he proposed new technical terms that now serve other authors well. Thus, Adamson (2007: 69–75) begins a chapter on skill with a discussion of Pye's terminology (see also Frayling 2011: 63–82). Pye chose to use the burdened words 'craft' and 'skill' as little as possible: the first of his precise alternatives is 'the workmanship of risk'. This means facture that involves dexterity and care. The workmanship of risk is contrasted with 'the workmanship of certainty', making that is jigged or guided (Pye [1968] 1985: 4–12). Risk and certainty do not describe good or bad ways of doing something. They describe how things are made. For this reason, they can be employed with clarity, unlike 'handmade' and 'manufactured', words that come freighted with prejudice. Pye ([1968] 1985: 13–19) proposed two further terms to describe the desired outcome of making, 'rough workmanship' and 'regulated workmanship'. Regulated workmanship is the accurate disposal of design, 'conveyed … by drawings and by specification' (ibid.: 21), generally aided by jigs or guides. The aim of regulated workmanship is to realize an idea with the minimum of tolerance. By contrast, rough workmanship is an approximation. Again, these two terms are relative, and again, they are neither good nor bad. They describe an attitude to instruction.

In its invocation of design, the idea of regulated workmanship describes the priority of vision over matter. Pye is quite clear on this point. It is expressed in the title of *The Nature and Art of Workmanship*'s opening chapter: 'Design proposes. Workmanship disposes' (ibid.: 1–3). To adapt Ingold's (2013: 20) phrase, design is a determinate end conceived in advance. It is an optical mode of making. Regulated workmanship is the process by which the design vision is disposed, the means by which matter assumes its form. But rather than being the end of the conversation, this rather lapidary division of labour into proposal and disposal opens the book. What follows is a rare attempt to understand the present tense of making, and how this stands in relation to the future tense of design. The crucial point is that workmanship indeed cannot be conveyed by words or by drawing, in advance. It is for this reason that workmanship merits its own book. Its strength is that it is pragmatically focused on immediate interactions between hand, tool and material, and the effects of different kinds of making. But if this book is read in isolation, the reader is left with a partial picture.

It might be thought that his earlier and less regarded work, *The Nature of Design* (1964), would concentrate on intended function. The reader is disabused of this notion from the start: 'What is design? … Most of the nonsense probably starts at the point where people begin talking about function as though it were something objective: something of which it could be said that it belonged to a thing' (Pye [1964] 1972: 7). If the function of the object does not belong to it, then how might it be prescribed by design?

> When you put energy into a system you can never choose what kind of changes shall take place and what kind of results shall remain. God has already decided those things. All you can do, and that only within limits, is to regulate the amounts of the various changes. This you do by design, which is the business of ensuring that at least you get the change you want along with all the others which you don't such as heat, noise, wear and the rest. It is as though the world operated on the principle of 'truck'. If you want some of this then you must take some of that as well, even if you do not want it.
>
> (Ibid.: 15)

Pye was critical of design methods that considered objects separately rather than as components in a larger system. The book is full of verbal and photographic images of systems of different scales. On the cover of my edition, there is a photograph of a regulator clock (Benjamin Vulliamy, *c.* 1780). Later on, Pye describes a particularly important slide-rest and leadscrew, made by Henry Maudslay in the early 1800s.[5] It is 'one of the most important determining systems and the parent of innumerable others' (Pye [1964] 1972: 54), an ancestor of the modern machine mill. All of these particular systems form their own internal economy, but in turn are part of a wider economy that contains all things: 'No system is self-contained. Every device is a subsidiary part of a more extended system (which must contain among its other components, man)' (ibid.: 31). If we consider Pye's recognition of 'truck' and his belief that the operative and the designer are both part of a dynamic system of flow, we might need to read his work on workmanship in a different light. Indeed, these books are larded with a strange sense of failure. Consider these quotations, one from each book:

> Nothing we design ever really works. ... The aircraft falls out of the sky or rams the earth full tilt and kills the people. It has to be tended like a new born babe.
>
> (Ibid.: 10)

> All workmanship is approximation. There are in the world of manufacture, and not only in that of metaphysics, certain Ideas of which the things we make are necessarily imperfect copies.
>
> (Pye [1968] 1985: 13)

Pye's theoretical contribution can be summarized as twofold. There is the very well-recognized contribution to our technical understanding of making. His categories of free and guided workmanship are still frequently referred to in the literature as precise critical alternatives to ideas of skill. His second contribution is much less well recognized. This is the idea that all human creative endeavour happens in a system of parts, people and material; and that, at a higher scale than this, each system is nested within a larger economy of materials and energy, long after and long before the act of making itself.

David Pye, in practice

The advantage of studying Pye is that we can do better than read his books against each other and think in the space between them. The evidence of Pye's own practice as a craftsperson can be used as well. Treating his fluting engine in this way forms a well-defined case for testing and developing his ideas. It is a very good example of a tool that forms a defined system, but which is also 'a subsidiary part of a more extended system' (Pye [1964] 1972: 31). The particular operation of this system is described below. It has two registers of time: the making of a single dish; and the longer evolution of the system, developing both the engine and the type-form of its products.

Pye describes the operation of the engine in a Crafts Council publication (1986: 43–9). It is a unique machine. When a bowl is turned on a conventional lathe, the woodturner works around the axis of rotation. The tooling appears as the spiral groove on a vinyl record, organized around the centre of the bowl. By contrast, Pye's innovation enables the operator to work at ninety degrees to the axis of rotation. In this case, the tool marks radiate in a linear fashion from the centre of the bowl. This can be seen in the surface of the rosewood dish. An understanding of the engine's operation is essential for understanding his relationship with the machine (Figure 7.2). The wooden 'blank' from which the bowl is to be carved is fixed to a rather stiff turntable. This can be rotated, as well as raised and lowered. The cutting blade of the fluting engine is a hook-shaped gouge. This is attached by a pivot to an A-frame that sits above the turntable. A lever operates this. As this is pulled, the gouge describes an arc towards, and through, the wooden blank. It carves an elliptical groove from the wood. At the conclusion of each pass, the operator rotates the turntable by a few degrees, and then repeats the process until a circle has been turned. The turntable can then be raised slightly, and the operation repeated. This enables the operator to shave successively deep grooves from the blank until the bowl is finished. Without adjustment, this would produce a symmetrical product, but Pye describes how complexity is introduced. The gouge can sit at an angle to the blank, and thereby carve an arc that moves through two axes. Alternatively, an oval-turner's chuck can be used so that the movement of the turntable describes an ellipse rather than a circle.[6] All of these processes open up new possibilities of form. None of these processes were previously obtainable for woodturners, and the forms would be unachievable unless carved entirely by hand. Pye (1986: 50) made at least one other fluting engine, a smaller horizontal version, adapted for carving patterns on the lids of wooden boxes.

Figure 7.2 David Pye, *Fluting Engine*, 1949–50, in its 1980s iteration. Courtesy of E. & J. Pye.

It is clear, from observing the fluting engine at first hand, that it is a system of interoperating parts. Taking our lead from Pye himself, let us speculate about his role as the human component in this system. There are some decisions that must be taken in advance of the equipment being used. The operator must choose a billet of wood and roughly shape the bowl with a 'heavy spout-adze' (ibid.: 46). The size of the billet creates an upper limit for the dimensions of the bowl. Rather than being concerned with the form of the bowl, the object at this stage is to 'show the figure of the wood – the pattern of the grain – to the best advantage, and to avoid defects in it which look unsightly or may weaken the thing to be made' (ibid.: 31). The gross form of the bowl is determined by the configuration given to the machine. By varying the relative positions of the cutter's pivot and the turntable centre, and by altering the cutter's radius and the angle of the plane through which it moves, a large variation of form is possible. If the oval-turner's chuck is used, then this increases it further. These parameters are fixed before the engine is used, and these prescribe a narrow set of potential outcomes. The form is latent in this configuration. The operation of the machine will translate this latent form from the engine to the material. Even in the production of one bowl, Pye and the engine are parts of an interrelated system. Furthermore, as the next section makes clear, the fluting engine is itself developing, through modifications made by Pye, and so the production of any individual dish should be regarded as a stage in the ongoing ontogenesis of the broader system of production. This procedural modification is instructive for contemporary digital craft.

Working with engines

Pye's fluting engine has a unique relationship with him, as the inventor of the tool and – in any meaningful sense – its only user.[7] Nevertheless, Pye's example has recently been emulated. An American craftsperson is in the process of making his own fluting engine after photographs of Pye's original (Wumkes 2017–18). But it is in relation to digital craft that the most interesting examples can be found. Even though he was writing in the 1960s, both of Pye's books describe the new technological horizon of numerical control:

> By numerical control certain designs can be translated (not interpreted) and 'told' directly to a machine tool so that a prototype or tool can be made without any care, judgement or dexterity being exercised at this stage. Ultimately automation may dispense with the operative altogether; but hardly the workman, who will presumably remain indispensable to it somewhere, even if numerical control advances to the point that a set of machines, given a suitable programme, can design and make another without the workman intervening at all.
>
> (Pye [1968] 1985: 26; see also Pye [1964] 1972: 55)

Pye's work is situated at the cusp of this technological revolution. It is perhaps not surprising that artists and designers have returned to it as a provocation, because it has a register of 'craft' but is mediated by machine. In 2011, the American designer and academic Zeke Leonard (2011) wrote about this and later created his own fluted bowl, *Homage to David Pye*, carved using a CNC routing machine (Figure 7.3). A few years later, the British design academic David Grimshaw made another *Homage to David Pye*, again using CNC technology (Kettle et al. 2014). Grimshaw says he was provoked by Pye's avowal that he (cited in Grimshaw 2017: 3737) would have been 'fascinated to see the sort of things I make being turned out by a computer … [but] I bet I could do it better than the computer all the same'.[8] Grimshaw (2017, n.d.) followed his first reimagining with a more formally complex object, a form that it would not be possible to make using Pye's process, but which nevertheless still quotes from the characteristic internal fluting of Pye's dishes (Figure 7.4).[9] Leonard and Grimshaw are both concerned with how and whether craft knowledge is manifested in their work. Leonard (2011) describes his homage as 'a craft object', but comments that 'something is lost in this particular amount of remove

Figure 7.3 Zeke Leonard, *Homage to David Pye*, *c.* 2010. Salvaged long-leaf pine, 229 × 229 × 51 mm. © Zeke Leonard.

between the maker and the material. … Having that interaction modulated by a computer screen … feels too distant for me.' Leonard's comment is consistent with what Grimshaw (2017: 3736) sees as a general tendency towards deskilling in design education: 'I believe that the development of the design process from the physical to the digital has had some undesirable consequences and the current emphasis on developing design work using screen based CAD [computer-aided design] is detaching designers from the physical experience of material making.' In response to this challenge, Grimshaw's project interprets Pye's bowl forms, using a process where 'the final physical characteristics are not fully determined within the CAD modelling phase' (ibid.: 3735). Instead these characteristics are determined through an iterative heuristic process, in which the precise interaction of wood grain and cutting blade is observed, and the stepover of the milling blade adjusted accordingly.[10] Thus Grimshaw opens up a gap in an otherwise rigidly regulated process. These breaks of free judgement allow him to apply 'craft making knowledge within the digital machining process' (ibid.: 3735) – to use Pye's ([1968] 1985: 26) terminology, in order to exercise his 'judgement or dexterity' – but as will be seen, the process is less continuous and less embodied than the example of Pye's own fluting engine.

Pye wrote relatively little about his own work, so it is not known how his fluting engine was conceived – whether he had a notion of a bowl in mind and

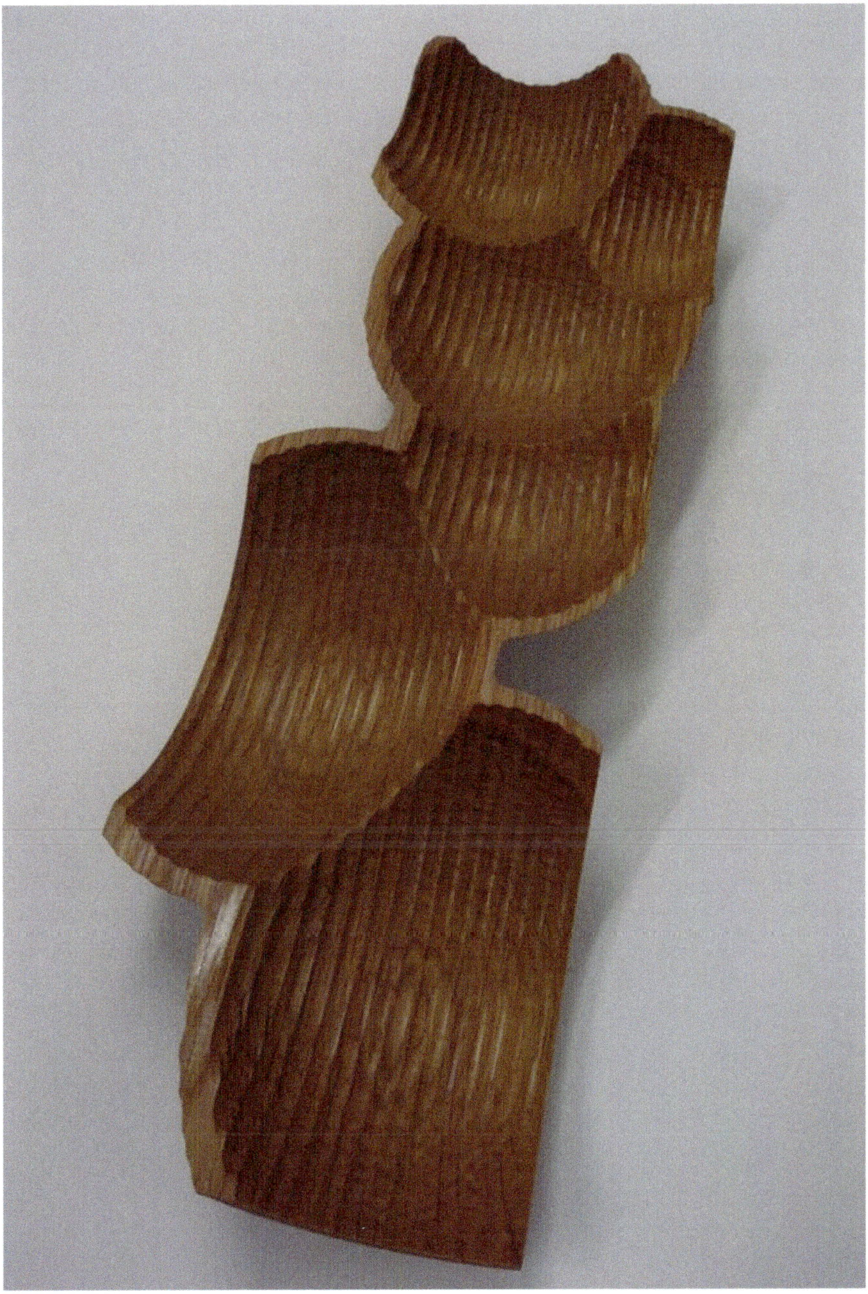

Figure 7.4 David Grimshaw, *Linked Elliptical Bowl*, *c.* 2016. Wood. © David Grimshaw.

wanted a tool to make it, or whether the idea of a scraping tool came first and its application followed. However, there are clues that show how Pye's work progressed. The first of these is that his engine, and so his bowls, became more complex over time. In 1986, he described adaptations that he had made to the machine over the preceding ten years. One was to swing the gouge in a plane diagonal to the turntable, enabling the creation of spiral flutes. The other was the introduction of the oval-turner's chuck (Pye 1986: 46). Before 1976, all of his bowls' depressions were, perforce, circular. At some point between 1976 and 1986, oval forms became possible. The second clue to how Pye developed his work is a short comment that he made about the objects of his second, smaller fluting engine. This device is for the engraving of wooden box lids:

> The technique for engraving … lids is simply an adaptation of the bowl fluting technique and I developed it about eight or ten years ago. It is far less adaptable than the technique of the Ornamental Turning Lathe (though probably much quicker) but it is still capable of producing a surprising variety of patterns – some of them very nasty!
>
> (Ibid.: 46)[11]

This is an admission of surprise and discovery. From the evidence of Pye's account, his iteration of bowls and dishes over four decades is a slow improvisation. Following Pye's definitions, this can be thought of as 'the workmanship of slow risk'. Rather than his projecting his vision of form through the engine and on to the matter of the wood, a repertoire of forms can be seen evolving. Thus we can see the broad typology of the rosewood dish that we are concerned with here, repeated in approximation and variation, sometimes with handles, sometimes without, and in different kinds of wood.[12] If we return, then, to Pye's idea of the human component, it is clear that this component is not simply a command centre, a programmer issuing instructions; the process is a kind of slow experimental play or improvisation. Rather than thinking about Pye's work only as the production of discrete bowls, it is more productive to conceive of this as a process of discovering the affordances of the system, of finding out what works and what does not. This creates a cognitive field, distributed through the technologies and materials that form the system of making and without which it is to be doubted whether these new forms would be created.[13]

So far the discussion has been limited to the determination of gross form, and the setting up of the machine – that is to say, everything that happens before the carving action of the engine starts – though it should be clear by now that the making process never really starts or stops. Let us turn our attention to that part in the process where a blank is in place and the machine is being used. The operator takes up the lever, to which the gouge is attached by a bell crank, and pulls it down, pushing the gouge through an arc, and shaving a sliver of wood from the blank. The turntable is then rotated by a small degree, and the operation is repeated. Eventually a whole circle is turned, and the bed is fractionally raised, allowing another pass to be made, carving a deeper depression. Incrementally, the bowl is formed. Each stage in the process is determined by the step before. We might think again of a modern process such as rapid prototyping, which is like a kind of sedimentation. But while it is running, Pye's engine has a human component that is following and modulating the flow of information within the system of production. As Pye explains, the fluting engine is only moderately jigged: 'I have never cared to put a dividing plate on the turntable so that all the flutes could be indexed to an equal width. It could be done, but would mean that the inside of the bowl was all regulated. I have preferred the element of freedom introduced by spacing the flutes by eye and hand' (1986: 46). Regulated workmanship is the accurate disposal of design 'conveyed ... by drawings and by specification' (Pye [1968] 1985: 21), generally aided by jigs or guides. In spacing the grooves 'by eye and hand', Pye is inside the adaptive operation of the engine and improvising through the process. While the broad shape of the dish and the discovery of pattern belong to a longer duration of evolution, the fine detail evolves while the dish is carved. This effect is clear on the surface of the rosewood dish. It looks regular, but – and especially once one knows how the engine works – it reveals the evidence of this incremental process. By looking at the surface of the rosewood dish, we can see that each pass of the gouge carves a flute, and that each flute forms a structuring basis for subsequent judgements and subsequent flutes. In the same way, each dish forms a structuring basis for the ongoing modification of the engine itself. It is the surface of the dish that tells us most about its concrete realization, and that gives most insight into Pye's relationship with his unique technology. Pye is embroiled in both systems of production: the production of an individual dish, and the evolution of the engine.

The French philosopher Gilbert Simondon describes this kind of process as 'transduction':

> This term denotes a process – be it physical, biological, mental or social – in which an activity gradually sets itself in motion, propagating within a given area, through a structuration of the different zones of the area over which it operates. ... The simplest image of the transductive process is furnished if one thinks of a crystal, beginning as a tiny seed, which grows and extends itself in all directions in its mother-water. Each layer of molecules that has already been constituted serves as the structuring basis for the layer that is being formed next, and the result is an amplifying reticular structure. The transductive process is thus an individuation in progress.
>
> (Simondon [1958] 1992: 313)

In this extract, Simondon describes individuation as movement from a pre-individual state in which being is immanent or virtual but not actual. The implication here is that the creative process is a repeated movement between individuation and disindividuation. In particular, Simondon's project exposes the 'technological insufficiency of the matter-form model' of ontogenesis (Deleuze and Guattari 1987: 408), which, in the context of this chapter, we can understand as the future tense of design ideation. Pye's practice involving the fluting engine is a good example of transduction in Simondon's terms, especially when we follow Pye's lead in locating the maker *within* the system of production: the latent potential of the system of making is distributed throughout all components of the system, from the billet of wood to the engine and the maker's own embodied material knowledge.

Pye's book on workmanship contains a typically lapidary description of a self-jigging tool. This provides a verbal image of a transductive process, just as much as his rosewood dish provides a concrete example: 'Many tools are partly self-jigging. The adze is, for one. The whole secret of using it accurately is that the curved back of the descending adze strikes tangentially on the flat surface left by the previous stroke – which becomes a partial jig – and rides along it so that the new stroke more or less continues the plane of its predecessor' (Pye [1968] 1985: 18). Pye's example of transduction is characteristically physical. An adze is a tool that is handheld, and it provides substantial feedback to the user of the tool, transmitted through its handle to the operator's hand and body. From a contemporary point of view, Pye's engine also looks heavy and clunky: something that must be negotiated through the human body as a part of the system. The two contemporary responses to Pye's work that I have mentioned sit in a somewhat different relation to the bodies of their makers. Leonard (2011) describes the loss of feeling involved in having 'interaction modulated by a computer screen' as opposed to the 'butt of a chisel'. Grimshaw describes a process that opens up to his judgement in stages, but that otherwise proceeds without the continuous embodied care – the slow risk – of Pye's fluting engine. By contrast, when Pye entered the system of production that we have been concerned with here, he must have felt considerable physical feedback as the hook gouge carved the wood. His daughter describes seeing as a child the shavings left by her father's work. Rather than being smooth, continuous lengths, they formed tight bunches of short arcs joined at the tip, leading to the family renaming the fluting engine the 'banana machine'.[14] These bunches are evidence of the effortful actions of the gouge through the wood. This feedback is concrete touch, as opposed to the much-discussed haptic perception – the virtual touch – of recent debate (see for example Ingold 2017; Bruno 2014). This is not to suggest that handwork or machine work is more material than its modern and digital equivalents, but there is an important difference. Digital technology demands less from the bodies of its users. But the fluting engine's operator must have had to contend with its materiality. Considering the likely effort involved in using the fluting engine, I suggest that it is not just Pye's mind that is extended through the engine, but his body too. We should consider his hand and engine together as the location of Pye's own embodied stylistic knowledge, his personal handwriting. Embodied knowledge is the kind of knowing-how-to-act that describes riding a bicycle or steering a ship. It also describes forms of cultural production, such as figure skating, drawing and carving a bowl. As the art historian and curator Lionel Lambourne (1986: 23) puts it: 'Handling a Pye bowl is like being at an ice rink immediately after the performance of a great skater, when we can trace the perfectly calculated, exactly controlled choreography, punctuated by a geometrically perfect "death spin", judged to a millimetre, the precisely considered effect of the inner and outer edge of the skate leaving a sharply engraved pattern.' This same author has described the 'unforced evolution of the "handwriting" of [Pye's] own style' (ibid.: 21). All of these actions are negotiated in the concrete realm, requiring the actual touch of the human component. This is what we see when we examine the surface of the rosewood dish.[15]

To summarize: in relation to Pye's practice we have described two types of system. The first is the durational system by which the engine, the broad morphology of the bowls, and Pye, as the human component, evolved. The second is the operational system at each specific point of time, in which individual pieces are made. In both types, the process is transductive. Therefore, it would be a mistake to imagine Pye projecting his design vision through the engine. Instead, we should understand the system as a process of negotiation and informed discovery. Of course, in practice, the design–workmanship divide does happen. There are designers who do not make. There are pattern-makers who do not design. Instructions can be issued, in words and diagrams, with a reasonable expectation that something will be made conforming to the command. But there are two aspects of Pye's work that must be taken into account in considering his theoretical legacy. The first of these is a critical sense of failure, the diminution of 'determinate ends conceived in advance': 'Everything we design and make is an improvisation, a lash-up, something inept and provisional. We live like castaways' (Pye [1964] 1972: 10). The second is his insistence that all action is related. No gesture, no thought, no movement takes place on its own. Everything that is made is contained in a system. Every system is contained within a larger system (ibid.: 31). The actions of the designer and the workman are restricted to making adjustments within a system of flow, from within it, and as a part of it. Pye's practice, in which he plays the twin role of designer and workman, forms an excellent illustration of this process.

Conclusion

This chapter has examined the surface of a rosewood dish by David Pye and its indexical relation to its system of production. The material and stylistic qualities of the surface of the dish give insight into the particular system of making and the role played by the maker's experiential and embodied knowledge, dexterity and care throughout the making process. This synchronic system is also a 'subsidiary part of a more extended system' of adaptive change (Pye [1964] 1972: 31). Here there is a reciprocal relationship between the gross morphology of the dishes and the increasing complexity of the engine. This diachronic system is marked and enabled by discovery and surprise, which is described, in a modification of Pye's terminology, as the workmanship of slow risk.

Pye's unique technology is situated at the cusp of the technological revolution of CNC manufacture. As this chapter describes, it has been a creative provocation to contemporary makers. Whereas Leonard's homage forms a somewhat mournful resignation to the separation of maker from material that is the default condition of computer-mediated practices, Grimshaw has designed a system of production that is iteratively opened to his material judgement and craft knowledge. In this way, Grimshaw's project is exemplary of the 'new kind of mastery of the hand in the iterative processes' of computer-aided manufacture (Harrod 2018: 12). It will be interesting to see how it develops from this point, and whether there is the same quality of haptic improvisation in the slow evolution of his project as can be traced through the evolution of Pye's work. Nevertheless for the present moment, there is a marked difference between the continuous physical contact of Pye's making and the slow risk of its evolution, and the iterative modification of jigged making that constitutes Grimshaw's project. Grimshaw (2017: 3736) has himself drawn attention to the etiolation of material skill in design education and the consequent loss to industry. His project is a response to this challenge. The close look at Pye's work that has been the subject of this chapter puts a sharper focus on this observation. The lesson of Pye's practice is to ask better questions of 'digital craft', and to be more sensitive to the difference between the haptic touch of digital creativity and the concrete touch of material palpation.

Notes

1 For instance, Peter Dormer (1997) presents a critical summary of Pye's writing on design and 'workmanship', but without reference to Pye's writing on *Ships* (1950), and – quite deliberately – without reference to the object of Pye's craft (Dormer 1997: 71). Dormer's partial focus misrepresents Pye's ideas.

2 It is notable that the Crafts Council (2020) avoids offering a definition of craft on their website.

3 Among the most esteemed are Bernard Leach, Yanagi Sōetsu and Alison Britton; for more examples, see 'Statements of Practice' sections in *The Journal of Modern Craft*. However, examples of craft writer-makers are far outweighed by instances of academic authors who are without the authority of making.

4 It is sometimes asserted (see Dormer 1997: 71) that Pye wrote two books on design and one on making. This is misleading. *The Nature and Aesthetics of Design* (1978) is a reprint of the earlier work *The Nature of Design* (1964), with the addition of some closing chapters on aesthetics. These have the quality of an afterthought.

5 A slide-rest is an adjustable device for holding tools on a lathe. A leadscrew moves the lathe's carriage. Used together, these appliances can regulate the production of spiral flutes.

6 An oval-turner's chuck is a device that regulates the lateral movement of the centre of rotation in lathes and similar machines.

7 This was certainly the case in 1986 (Pye 1986: 43). Sometime after Pye's death, his cousin, the sculptor William Pye, attempted to better understand the operation of the fluting engine.

8 See Frayling (2011: 93–108) for a full transcript of this conversation.

9 This work was exhibited at the Crafts Study Centre, University for the Creative Arts, Farnham, UK, 13 November–15 December 2018.

10 A 'stepover' is the amount of overlap provided for by each pass of a cutting blade in CNC milling. Typically, milling blades are rounded, and so a smaller stepover produces a more finely scalloped surface. Routed wood shows a more consistent grain when it is more finely scalloped.

11 An ornamental turning lathe is a complex machine tool for carving, for instance, barley twist candlesticks.

12 I am grateful to an anonymous peer reviewer for drawing my attention to Pye's concern with the notion of diversity in made things. This is a quality of variation that is lost in the workmanship of certainty, and which is 'most valuable in the contrast and tension between regulation and diversity' (Pye [1968] 1985: 36). The fluting engine could be considered an attempt to develop 'limited risk' in manufacture. However, in this example, the regulation of risk changes over time, as a result of the improvisatory relationship between dish, maker and tool, and the corresponding evolution of the fluting engine. Moreover, the evolution of gross form in Pye's work is negotiated without predetermining its end point, and in concert with this slowly changing regulation. Thus, diversity emerges in Pye's practice as gradual change in the potential limits of production. The duration of practice and the improvisatory feedback between change and regulation is what I seek to emphasize here: hence 'slow risk'.

13 For a contemporary example of a similar process of form evolution through technological innovation, see Jerhoen Verhoeven's *Cinderella Table* (2004; Johnston 2015: 258–61; Openshaw 2015: 263–5; V&A 2016).

14 Elizabeth Pye, conversation with the author, 2019.

15 I am grateful to an anonymous peer reviewer for the observation that further analysis along the lines of Gell ([1985] 2013) may reveal more regarding the link between the surface of the bowl and Pye's distributed personhood beyond the confines of the third dimension.

List of Figures

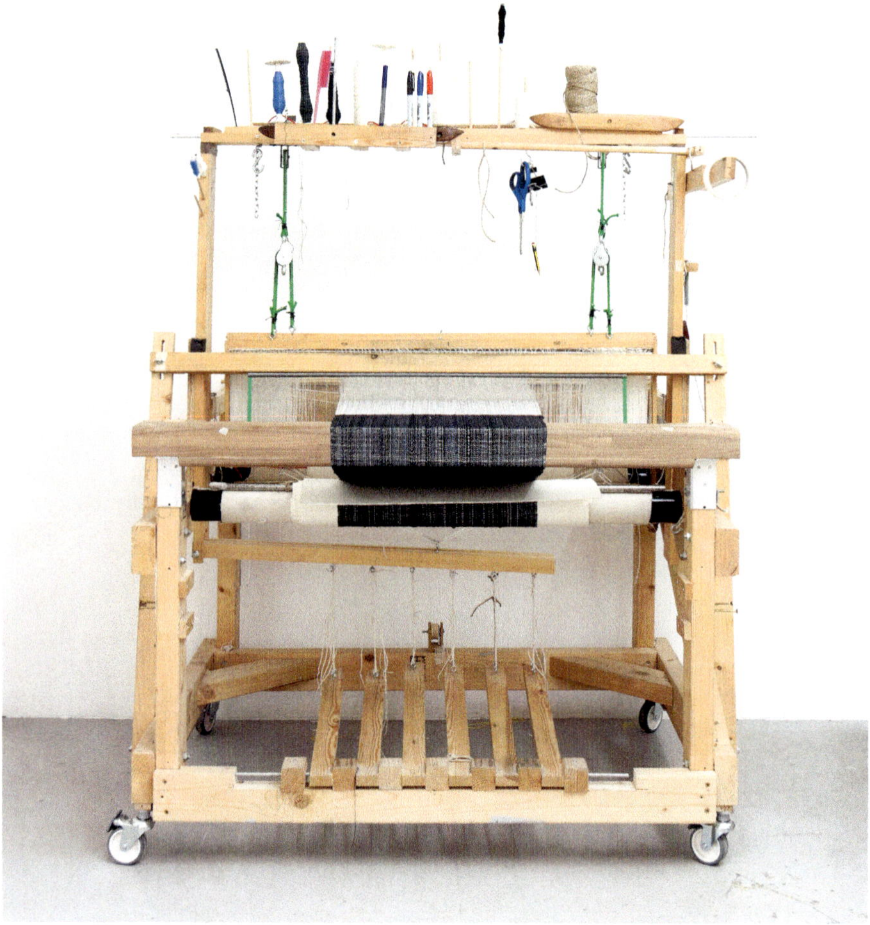

Figure 8.1 *Loom*, *c.* 2016–18. Wood, metal, plastic, canvas, cord, 165 × 120 × 110 cm. © Max Mosscrop.

Chapter 8
Journal (2016–2018): A Conversation on Looms, Cloth and Weaving

Max Mosscrop and
Benedict Carpenter van Barthold

Max Mosscrop is a British artist. He was born in Lancashire, England, a part of the world that has a long tradition of textiles manufacture. Having initially studied architecture, Mosscrop shifted his attention to painting and most recently to weaving. He was joined by the sculptor Benedict Carpenter van Barthold to talk about his new work, *Journal (2016–2018)*, at his studio in south London in January 2019.

Benedict Carpenter van Barthold: You used to make paintings in the conventional sense of a layer of paint on stretched canvas. Painted canvas is a very privileged sort of surface in a way that woven cloth is not. What drew you to weaving?

Max Mosscrop: Initially, weaving offered a solution to a formal problem in painting. Around 2014, I was making a series of paintings by overlapping long vertical and horizontal brushstrokes using egg tempera on gesso panels. I wanted to construct a surface with paint but felt hampered by the need to make a panel with a specific dimension, before I could make the painting. Weaving seemed like a logical step to liberate the constructed surface from its support. I began by making some things with wicker, like flattened baskets, and then I experimented with a very rudimentary frame loom. The direction seemed promising. In the past I've enjoyed making bits of equipment to aid the painting process, and I was excited at the prospect of all the paraphernalia that goes with weaving – the physicality of that and of the weaving process itself. The structure of woven fabric – the simple interlacing of threads – is beautifully economic, yet there is the potential for endless variation. This structure is a direct result of the making process, which can be traced in the surface of the cloth. There is an integrity there that I find compelling.

BCvB: You have previously told me that your grandfather worked in a weaving mill. Is this relevant to what you are doing now?

MM: My grandfather was twelve when he started work in the weaving mill outside Blackburn, Lancashire in 1902. This points to a certain historical context. At school we learned how industrialization spread from the 'satanic mills' of the Lancashire textile industry. Cotton was Britain's biggest manufacturing industry in the early nineteenth century, but most of the weaving was still done on handlooms right up to the 1830s, even though other stages such as spinning had become fully mechanized (Bythell 1969: 5, 28). So it's interesting to think about handweaving against the background of this incredible explosion in production and invention, and all that went with that, both good and bad. Handweaving now has the status of a 'craft' only because it has been superseded by industrial technology, but the basic principles of all weaving are the same. I am interested in the reciprocal relationship between the loom and the cloth, each as the product of the other, and both as physicalizations of the same ideas and procedures.

BCvB: Could you talk me through the cloths that you have made?

MM: *Journal (2016–2018)* is a long series of cotton cloths, the title referring to the daily, incremental way in which they were made. They are 40 centimetres wide and vary in length, with a combined length of about 40 metres. I used black weft on white warp to emphasize the interlocking of threads. The black and white also suggests the shared etymology of 'text' and 'textile'. I'm posing a question about what kind of knowledge might be produced through weaving as a daily practice of inscription, as if writing in a journal. There is a convention in weaving called the 'sampler', which is a format for experimenting with evolving ideas in a single cloth, as a learning or a test of skill. *Journal* borrows this convention, and the cloths are articulated in lateral bands, each with a different weave structure and texture. I threaded the warp through the harness[1] in randomized irregular sequences, so that there is no repetition of pattern across the width of the surface. I made a long list of possible treadling sequences and then used dice to choose from the list. So each band is the outcome of an experiment – how will it be if I use a particular treadling sequence? – and together they form an archive of my own learning. The incremental nature of the weaving process means that each passage of weft corresponds to a particular moment in time. As such they form a record of my movement as a weaver.

BCvB: You could have used a ready-made loom, but you have made your own. Why?

Figure 8.2 *Journal (2016–2018)*, *c.* 2016. Process, film stills. © Max Mosscrop.

MM: Building my own loom was a way of incorporating weaving and its process more deeply into my practice. It also seemed like a good way of learning about the principles of weaving. I have now built a second, larger loom, to my own design, but for the first one, I followed instructions from a manual called *All About Weaving: A Comprehensive Guide to the Craft* by Clara Creager. It felt strange, to be an artist making something by following someone else's instructions. Not quite an act of appropriation, but a translation from a set of drawn and written instructions into a material object, according to the tools and materials that were available. The loom-building has become an ongoing part of my practice as I continue to mend and refine its operation, and there is a constant feedback between the woven cloth, the weaver and the loom. This wouldn't feel so possible if the loom wasn't my own in this way. The weaving has produced the loom just as much as the loom has produced the weaving.

BCvB: From what you have said, is the cloth never really separate or separable from the process of weaving and, by extension, the loom?

MM: The process of weaving is immersive. As a weaver you put yourself in the middle of all this material and equipment, and the boundaries between weaver, equipment and material get quite blurred. Let me explain about the operation of the loom in order to clarify this. The loom is a mechanism for organizing threads. It keeps the warp threads in order and under tension, stretched between rollers at the front and back of the loom. In the centre of the loom are the harness and beater[2] assemblies, where the cloth is formed through the perpendicular intersection of warp and weft. Each warp thread passes through a heddle,[3] and each heddle is held in one of four shafts. The shafts are connected to treadles by an arrangement of cords and pulleys.

Figure 8.3 *Journal (2016–2018)*, c. 2016. Process, film stills. © Max Mosscrop.

Pressing a treadle causes the raising of certain shafts and the lowering of the others, which also raises or lowers a selection of the warp threads. This forms a space between alternate warps – a shed – through which a weft thread can be inserted. The weft is thereby made to pass over some warp threads and under others, before being beaten into place at the fell[4] of the cloth. This operation is repeated, pressing a different treadle, creating a different shed, and so capturing the weft between the warps. The warp is set onto the loom at the beginning of the weaving process, becoming integrated into the loom's physical structure. Each time a weft is inserted in the shed, its significance is determined in relation to the warp threads and all the previously placed wefts. And once the cloth has been woven, the warp must be cut in order to remove it. From my perspective, the loom and the material it organizes are not separate but form parts of an assembly. The semi-formed emergent cloth is a key component of the apparatus.

With handweaving, the weaver is another component. I have to climb into my loom, becoming a part of it. If you compare the handloom with an industrial loom, you can see that the weaver serves two functions: providing power by treadling, throwing the shuttle and beating the weft; and acting as a channel for the information that determines the treadling sequence and weave structure. When I'm weaving, I'm always following a predetermined sequence of operations, and my actions are determined by the physical structure of the loom. I participate in this flow of information, as well as being a physical component of the loom. The cloth, the loom and the weaver are so tangled up with one another that I see them all as a single assemblage.

BCvB: You have talked about the information involved in weaving. There is a connection between weaving and computing, with the Jacquard loom being regarded as a forerunner of the modern computer. A loom is a machine that reads and writes digital information, the ones and zeroes of digital computing corresponding to the relative position – the 'significance' in your words – of a weft thread being either over or under a warp. Is this connection to modern digital technology, and the virtuality it affords, a useful point of reference in thinking about your work?

MM: Weaving is an informational process: the information needed to make a particular cloth is conventionally specified in a pattern draft. The weaver transforms the draft into bodily movements. When I press a treadle, the loom picks up a selection of threads, effectively performing hundreds of simultaneous decisions. In this sense the loom is like a computer, but with a loom like mine, everything is visible. You can see how it works, how a certain component or function affects the quality of the cloth as it's being woven. And because I made the loom, I feel that I can get inside and alter something that isn't working or could be made to work differently. You use the term 'virtuality'. I understand this to refer to a set of potentials latent in a system but not necessarily actualized. One can think of the loom as a physical embodiment of such a system, and a woven cloth as a particular actualization. For me, the potency of the loom as an object lies in this sense of possibility. My loom refers to the past, to the history of weaving, but it also contains this sense of the future.

BCvB: Let's think about the experience of handling information in weaving. How does the physical effort of weaving make itself visible in your cloths?

MM: Weaving is repetitive and rhythmical, and this rhythm is reflected in the surface of the cloth. You can consider the woven surface in terms of two kinds of information. First, there's the binary information that determines the woven structure. But there is another kind of information which is registered in qualitative irregularities rather than structure. Slight variations in warp tension, bobbin winding or beating pressure will appear in handwoven cloth as departures from uniformity. It is interesting to compare a pattern draft – a design intention – with a corresponding piece of woven cloth. The differences between the two are an index of the physicality of the process and an inscription of contingency. It is gratifying that such a disciplined method leaves so much room for chance.

BCvB: So, let's think about the connection between these two different forms of information. There is the digital information of weaving, one concrete instance of which is written into the structure of each particular cloth. And then there is a kind of virtual array of possible cloths, encoded in the potential of the loom itself and the weaver's experiential knowledge, and this array is intimated by both the loom and any particular cloth. The material and the immaterial here give access to each other. This reciprocal access is accentuated by a processual modulation of the digital information, leading to variable quality of resulting cloth. Thinking about the loom as a kind of prototypical computer, in this coming together of the material and the immaterial, is there something in the surface of your cloths that might help us to understand our everyday experience of digital technology, or to rethink some of our assumptions here?

MM: This is an interesting question, and my answer is quite speculative. As digital technologies have become more sophisticated, our interface with them has become more sensorial rather than textual. I'm thinking of the action of swiping or voice-commanding a smartphone, for example. Woven textile is an affective surface. We experience it as texture and pattern, each with a particular rhythm, density, frequency and direction, and we find these felt characteristics difficult to name. But if you look at woven textile closely, you see that it is made up of these simple intersections of one yarn with another. At the point of making, each intersection corresponded to a simple decision: over or under. So it's as if these simple logical ideas accumulate to form a very physical, affective experience. Perhaps something similar happens with our digital devices?

Notes

1 Arrangement of wooden or metal frames (shafts) from which heddles are suspended, and which can be raised or lowered. A heddle is a length of wire or string with an eye in the middle through which a warp thread is passed.

2 An assembly pivoted from the top or bottom of the loom, used to beat the weft into place.

3 Length of wire or string with an eye in the middle through which a warp thread is passed.

4 The point in the cloth being woven where the last laid weft thread lies.

List of Figures

Chapter 9
On Drawing: Transmission from the Lifeworld to Paper at Namdaemun Market, Seoul

Ray Lucas

Introduction

This chapter presents a series of survey drawings as a *graphic anthropology* of Namdaemun (Seoul) and Seomun (Daegu) markets in South Korea. In transmitting the complexity of these markets – which arrange themselves around display and surface, capable of rapid adaptation and transformation – into drawings on paper, it reveals the superficial nature of the market, rehabilitating the term as a positive quality. The approach has some of its roots in the anthropological concept of the lifeworld, which offers a more holistic idea of context than that offered by conventional architectural site studies. An alternative to the problematically enclosed ideas of 'culture' or 'society', the *lifeworld* (Jackson 2012: 7; Lucas 2020a: 2–4) is an intermixed and enmeshed condition where people and their environments co-produce one another: people adapt their surroundings to their needs, but that place also has an impact upon their lives. These lifeworlds typically lead to the development of particular ways of living that are unique to that place, and the understanding of these ways of accommodating and being accommodated is a central concern of anthropology. My work lies at the intersection of architecture and anthropology, making use of anthropological concepts in order to articulate the social nature of architecture and to find circumstances where architecture can move beyond its conventional bounds.

Lifeworlds posit that the very act of living and dwelling is a way of producing and articulating knowledge about the world: and I contend that this knowledge production can be usefully extended to architectural practices of understanding through drawing, model-making and building. Lifeworlds help us to understand that social life cannot be understood in isolation, but across a range of scales and activities.

Namdaemun Market in central Seoul is a general market selling a variety of goods from both permanent and temporary stalls. The market consists of a quarter of the city next to the Sungnyemun gate, bordering the commercial district of Myeongdong and major institutions such as the Bank of Korea Museum and Seoul Central Post Office. The market's operation varies from day to day, with the grid of streets between the blank and monolithic market buildings occupied by a variety of vendors. This chapter focuses on describing the street market, with its ad hoc improvisations, cooperative supervision, spillages from retail units and adaptations of modular carts.[1]

This chapter presents some of the methodological implications of a graphic anthropology of the market, where the approach has been to take the production of drawings, notations and other inscriptive practices seriously as a form of research. The strategy of multiple representations is at the heart of this process: where one form of drawing edits details out, another will pick these up and focus on them. By drawing and redrawing, notating and diagramming, a broader range of phenomena can be described. While individual drawings may appear to 'clean' the market up, this is not the intention; by reading across the entire set of drawings, the thickness of the description emerges. Isolating individual drawings directs the attention to details, while the set allows us to appreciate the complexity of the site. The idea that drawing constitutes a form of knowledge production is experiencing a resurgence, particularly within anthropology. Alternative forms of graphic anthropology are under development, many of which focus on narrative and sequential art more commonly found in graphic novels and comics. I present an alternative approach here,[2] founded in the conventions of architectural drawing. These conventions have developed for a number of pragmatic purposes, primarily the understanding of geometry and space. More often used speculatively in the design process, my approach here has been to use the architectural drawing conventions of orthographic and parallel projection as analytical tools, and as a way of placing the ephemeral constructions of Namdaemun Market into circulation as explicitly *architectural* phenomena.[3]

The process of drawing from the field is modelled here as transmission, a term problematized by Ingold (2007a: 115–16) as carrying information (largely) over time and allowing a spatial dislocation, as opposed to the lines of transport which carry information (mostly) across space. According to Ingold's account, both forms of relationship have a spatial and temporal aspect, communicated along lines. This allows me to suggest several things about the drawing process: that the preference for in-situ drawing establishes a false dichotomy with later studio renderings or, as I prefer, post-situ drawings based on a range of preliminary sketches, photographs taken as records and source material, and also based on memory and informed supposition or best guesses. This is part of Ingold's larger concept of wayfaring and wayfinding, whereby ideas of rationalized point-to-point transportation are supplanted by a process of discovery and continual feedback between the environment and its inhabitant.

Discussing survey drawings from antiquity to the modern period, Carolyn Yerkes (2017: 14–15) presents Renaissance and subsequent architectural surveys as a multifaceted intersection of in-situ drawing with other historic and contemporary sources. The survey is embedded as an important form of architectural knowledge production from this early stage in the profession's development: that knowledge gained from literature was a foundation, to be supplemented by first-hand observations which have the benefit of being driven by the agenda of the architect in question. The material contained in this chapter has its lineage in such surveys, informed by the problematization of such early tourism and the ongoing discussions regarding fieldwork in anthropology. Architects can advance the discipline through an understanding wrought from survey, producing a knowledge distinctly architectural, sitting at an intersection of the spatial, material and social. This project asks what the professional discipline of architecture can learn from the iteratively produced designs of market vendors, both in terms of functional and material efficiency, economically driven choices, and the social aspects of cooperation, competition, legitimacy and exchange.

The drawing process here is an extension of anthropological concepts of the co-production of environments and people: inscription is made possible by elongating moments experienced, dwelling within a collection of moments in order to meditate further on the material and geometric qualities of a space, the atmospheres of a place or the arrangements and relationships between elements. These are apprehended in brief moments and cast immediately into memory in the manner described by Bergson (1992: 100); our perceptions are not immediate but informed by all prior experiences that form our understanding of the world. Drawing is understood as a temporal process, finding ways to make the fleeting persistent and to embed these traces into the surface of paper or other similar supports.

One common assumption is that the drawing process makes for a deeply subjective experience, but architectural studio culture gives us some hints about how this might be overturned. Architectural studios often employ a house style, determining everything from the positioning of elements on the page, to the fonts used for lettering, the line weights and codes used to indicate materials and building components. Drawings also have an aspect of editing: selections are made, and these are essential to the drawing practice's status as a form of knowledge production. Elaborating the role of the *researcher as draughtsperson*, and considering that – given the use of a 'house style' of materials and line weights – another similarly trained architect or draughtsperson could easily produce the same drawings as me, reveals that, in a sense, my drawings are *not* autographic (Goodman 1976: 113). In the end, the discussion around subjectivity and objectivity gets us nowhere useful; it is a spectrum, and this work is positioned at a precise place and time, with the researcher present in the space. Objectivity is a measure not easily applied to anthropologically informed architectural research, nor should it be. That is not to say that it is expressionistic in nature: the protocol for the drawings is repeatable and verifiable.

Surfaces are both implicit and explicit in the series of drawings that follow. The first instance is the very act of making lines into a surface as a way of communicating the spaces of Namdaemun and other markets. In fine art practice, the surface is most often discussed as a *support* for the medium. This establishes a clear hierarchy in the materiality of drawing, relegating the surface to a secondary status. In the case of the main series of drawings, the support is carefully selected after trial and error with different surfaces. The board is then prepared carefully to allow a certain kind of drawing to be made: a grid of dots is applied with a printer to allow free-hand architectural drawings to retain a sense of accuracy and measure.[4] These arrangements make use of the materials: different types of paper and card supports with varying weights and textures, resistance and accommodation to the medium. These media are in turn made from graphite, clay, pigments and dyes, and are applied with brushes, nibs and pencil leads. Each of these choices carries a range of affordances and excludes others: the colour permitted by marker pens is difficult to control, so the precise lines of pencil and fine-liner work can get lost; watercolours allow for atmospheres to be communicated, but are unruly materials, more difficult to control every aspect of their interaction with absorbent paper.

The second set of surfaces are those of the markets. Translating these surfaces into drawings allows for an analysis of their qualities, and the drawing conventions inform that understanding greatly. The nature of surface is, itself, a pertinent research question. Architecture has a body of tacit knowledge regarding the surface treatment of buildings (explored by Leatherbarrow and Mostafavi 2002), largely formed by an understanding of classical Western traditions. The façade, in this narrative, is a communicative surface, informing the viewer of how forces are transmitted from the roof to the ground, and later as expressions of a building's function. Further discussions of surface lead us towards the work of environmental psychologist James Gibson ([1979] 1986: 307) and its later advancement by anthropologist Tim Ingold (2000: 209). Ingold maps the earth-as-surface to anthropological notions of *lifeworld* as an experiential alternative to global concepts of space: we are *in* the world rather than *on* it. In this understanding of surface, it is an essential component of environmental knowledge: surface has the role of mediating between the firm and immovable *substance* and the thick but traversable *medium*. Here, surface is a component of perception reconceived as an ecological phenomenon, a wider system that responds to how each animal dwells in the world: for moles and worms the earth is a *medium*, in contrast to its status as a substance for most other animals. For conventional architectural expressions, Gibson's ecological perception transfers quite logically, but is complicated by the furtive and temporal nature of the market. Examples include the thickness of a wall that accommodates a luggage vendor or convenience store, with sometimes as little as 300 millimetres allowing for a complex economic exchange to take place. The surfaces of the market stalls are present, repaired and iteratively redesigned, taped together and propped up, laid on the ground or hanging overhead to demarcate territory. Some of these surfaces are depicted and others are elided or edited out in order to make the drawings focused and legible, but the context is fundamentally one of surface modulation whereby interactions are framed and guided by the information surfaces are giving us. The sections that follow will discuss the ways in which inscriptive practices describe these complex surfaces, with each drawing and notational convention highlighting the pertinent features of each part of the market.

Serial vision and the politics of perspective

This selection of drawings describes a walk in Seomun Market in Daegu in 2014. Seomun is similar to Namdaemun in that it is a general market with a mixture of ad hoc stalls and urban blocks housing more formal market areas. Seomun is hemmed in by the construction of a new expressway and suffered major fires in 2005 and 2016. The drawings use a form of *serial vision* pioneered by the British architect and urban designer Gordon Cullen, who used sequential perspective drawings as part of his critique of post-war modernist townscape in the UK (Figure 9.1).[5] The method is presented by Cullen as neutral, but this is far from the case. The process of producing a suite of perspective drawings along a path reveals visual variety, and is a critique of the monotony identified by Cullen in modernist reconstruction. By using Italian towns and undamaged English towns as a model, Cullen is deliberately problematizing the materiality and form of contemporary architecture, and the method is designed around this preference. This does not devalue the approach, however: using it knowingly and recognizing the way in which it functions allows a broader range of expressions and critiques to be made through Cullen's method of serial vision.

This sequence of drawings describes the complex three-dimensional nature of Seomun market: the manner in which flights of stairs are occupied for display, parasols are used to define areas and platforms grant greater access to the upper levels to both pedestrians and vehicles. Taking the pages of an A5 sketchbook and drawing in situ, positions are taken at regular intervals along a walk through the site. The drawings themselves are larger than thumbnail sketches, but smaller than the work of contemporary urban sketching movements.[6] The drawings demonstrate an alternative kind of visual complexity, one that is temporary and

Figure 9.1 Serial perspective drawing sequence of Seomun Market, Daegu, *c.* 2012. Pencil in A5 Leuchtteurm Sketchbook. © Ray Lucas.

changeable: aspects Cullen would routinely edit out of his depictions as part of a political reclamation of the picturesque against modernism.[7] Cullen would routinely edit his perspective drawings to omit people and street furniture, leaving in landscaping and the enduring structures of the architecture. In our example, however, movable and temporary elements dominate the scene: it would be completely devoid of any life without these elements, so it is essential to include the market stalls, awnings and parasols, goods for sale and in storage, tarpaulins and bungee cords. The drawings can hardly be described as *picturesque* and respond well to the method employed: people are still omitted for clarity, but the clutter of real life is described. Visual variety is clearly present, but not in the romantic way proposed by Pevsner and Cullen.

Figure 9.2 Cross-section drawings of Namdaemun Market alleyways, Seoul, *c.* 2014–16. Pencil in A4 Leuchtteurm Sketchbook. © Ray Lucas.

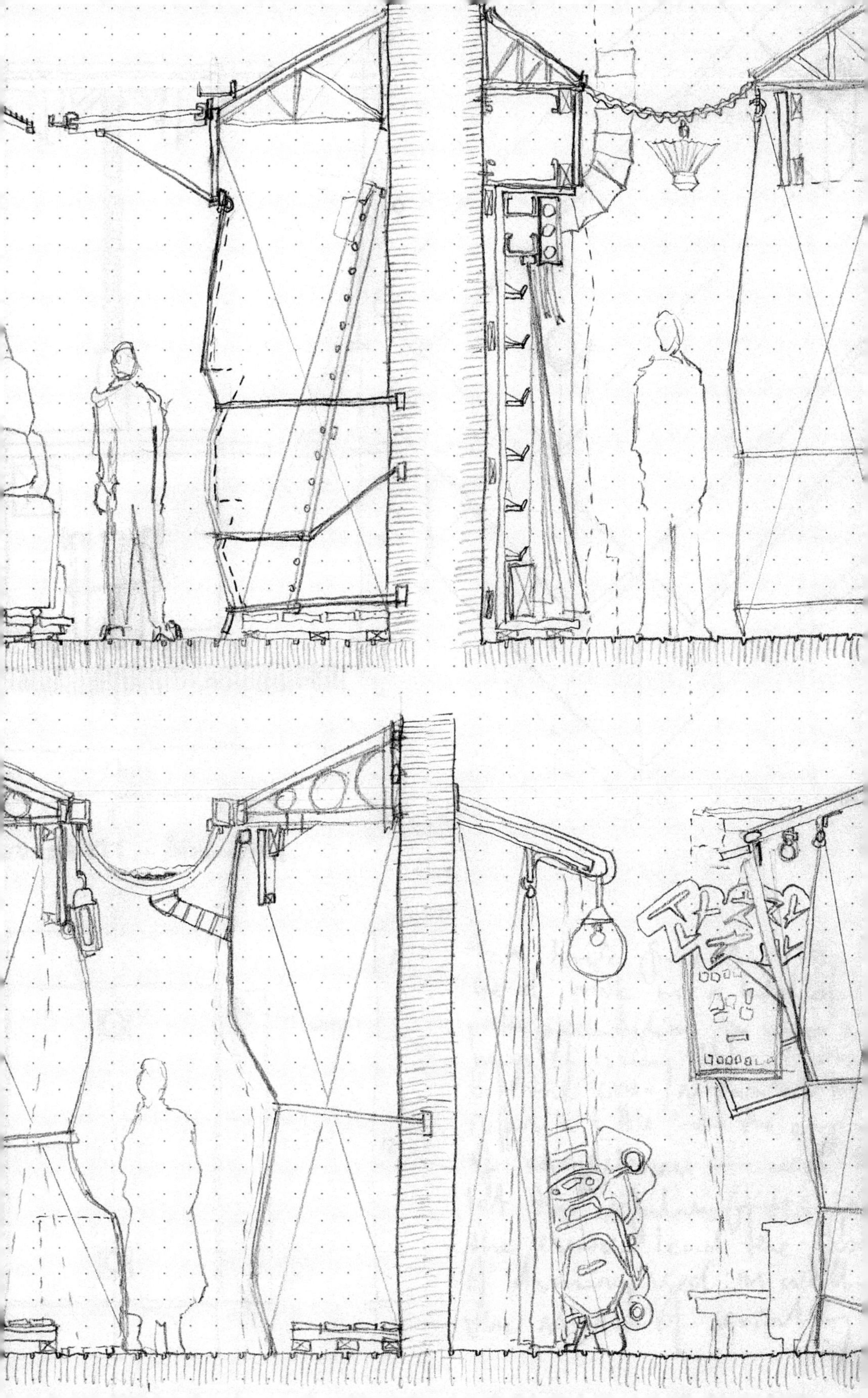

Thick surfaces in cross-section

Sections are a form of orthographic drawing arranged around a slice through the building or object in question.[8] Longitudinal sections and cross-sections are typically used to describe the direction of the cut, and architectural convention often shows a preference for the shorter cross-section presented in series (Figure 9.2).[9] The drawings shown here are of four alleyways at Namdaemun Market, spaces between the main buildings but away from the thoroughfares. These spaces are still busy and packed with activity, revealing a series of spaces for selling goods, storage strategies, restaurants which serve the needs of vendors rather than buyers, and servicing to support these activities. The drawings are made in a sketchbook with lightweight A4 paper, marked with a grid of grey dots. The lines of the drawing are made with a 2B 0.5 millimetre mechanical pencil, balancing precision with the ability to erase and rework the drawings as appropriate. Of particular interest is the thick surface of the alleyway, where a permitted depth is tacitly acknowledged by vendors, allowing the route to continue to function for foot and vehicle traffic (mainly motorbikes), while giving further opportunities for shelving and racking with goods for sale. This is covered over when not in use, with a thick sheet of tarpaulin fixed to hooks and secured with nylon cords or strapping. The surface is either busy with goods to attract the eye, or a bold textile, waxy and shiny, reflecting the artificial lighting needed in these tight spaces.

The section is particularly useful for depicting this thickness of a section, where walls have depth and can contain some function or other. Robin Evans, in his essay on alternative depictions of eighteenth-century interiors (also named ‘sections’), describes this as a *developed surface*, where decorative mouldings were carefully considered as parts of the design and therefore needed to be described in drawings. Sections, however they are arranged, then, have a common aim of describing this articulation of surface. The operation of the market activates the façade in a manner distinctive from the traditional understandings wrought of classical architecture, where orderly arrangements of construction materials express the forces distributed from the roof to the ground via walls and columns, around windows and doors. A traditional section shows the overhanging mouldings of classical features, the visual separation of the pediment from supporting columns and walls, with elevation showing background elements and how these relate.

These sections analyse the accumulations of adaptations made to these leftover spaces, a resource that is at a premium in the marketplace context.

By making the most of these neglected routes, activating them with functions adjacent to the main market, they support the overall ecology of Namdaemun. The density of occupation is shown by the scale figure, and the entirety of architecture's role in adapting an environment to make it more usable by people is demonstrated by the mess of restaurant extraction units, cost effective roofing, signage and lighting.

Elevation and the datum of the market cart

The mass-produced steel market carts provided by the city authorities are used in a wide variety of ways by vendors. The affordances offered by this simple structure are represented in a series of elevation studies, with graphite and red pencil depicting the cart and the goods for sale respectively.

The carts begin as compact units on castors, able to be chained together into long trains. These are then trailed around the market site in the evening by quad bikes and returned to a storage area. Once in place, the cart needs to be connected to the market's power supply, as the roof to the unit is mechanically powered. Once activated, the roof lifts slowly, and can be unfurled to provide a wider canopy. A similar surface area can also be unpacked at the counter level, providing a platform with storage in the main unit. These official carts are quite different in character to the informal ones: the materials are polished and clean rather than reused and ad hoc; the wheels are small, requiring a smooth and level surface to move over; the self-built stalls are either fixed in place permanently, or mounted on wheels capable of coping with a wider range of surfaces. In short, these carts are ergonomically designed to allow for a wide variety of uses, whereas the iteratively designed ad hoc carts are carefully designed for one purpose (Figure 9.3).

The drawings of the elevations show how the cart is used, by different vendors, with some common solutions including the use of the roof to store goods; hooks attached to various parts of the structure to hang mirrors or price tags; and stools, chairs and flooring set down to demarcate public and private zones around the stall. The cart can be used with a distinctive front and back; as a 360-degree platform for showing goods; as the base unit with cooking facilities for a much larger structure by stretching sheeting across an area to provide seating for a restaurant.

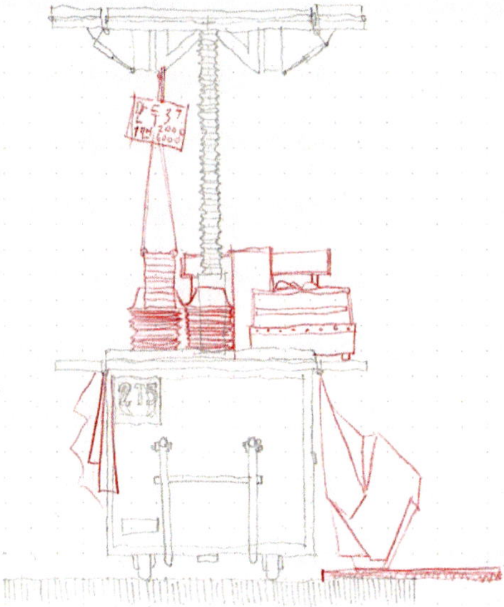
275

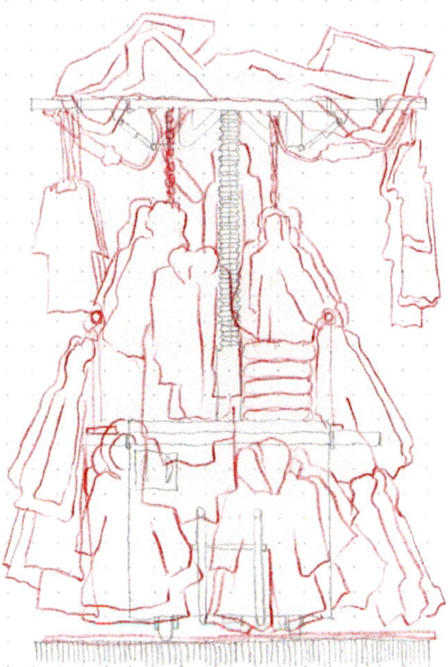

Plan and spatial relationships

The plan is a more commonly understood form of projection, sharing much with maps, as an overhead view describing the spatial relationships between discrete features. Plans in architecture are, of course, designed to have features relevant to the built environment. One of the best-known early examples is the Nolli Plan of Rome (*c.* 1748): a survey of the city that depicts its urban fabric. The most notable feature of the Nolli plan is that it explores the public and private space of the city through a simple but sophisticated device: the plan renders private spaces as solid, shaded blocks, while public squares are left as blank space. Further to this, the many churches of this important Christian centre are rendered as detailed interior plans, showing colonnades, arches, naves and aisles.

Alberto Perez-Gomez discusses the plan as distinct from orthographic projection or *orthographia*, reserving this definition for elevations. Plans are instead understood as *ichnographia*; Perez-Gomez (2012: 17) argues that the tripartite architectural *idea* is completed by *sciaographia* – misinterpreted according to Perez-Gomez as *scenographia*, or perspective. Our focus here is the *ichnographia* of the plan; in his celebration of paper and its qualities, Mario Carpo presents the plan as a form of divination (2011: 28–9), focusing on the nature of the convention in describing dimensions, to designate what is to be made.

In this regard, several sketch plans of Namdaemun and Seomun markets are presented below. The most important distinction is between the more durable (if not permanent) buildings and the market stalls that have a rather more complex duration. By simply colour-coding these sketch plans, the extent of the more informal aspects of the market are revealed. There is something interesting, however, in the production of such drawings. While in Daegu, for example, drawing plans in the street aroused some suspicion from vendors, as there are implications of legitimacy and authority around this form of territorial inscription. The sensitivity around mapping the spatial arrangement exposes some of the negotiation at the heart of the site of exchange: the drawing, even the sketch plan, is potentially powerful and that power must be exerted with care.

Sensory notations and representing the non-visual

Architectural representation is known to privilege the visual and geometric aspects of space, having difficulties in describing other senses. This often means

Figure 9.3 Elevation drawings of modular Namdaemun Market carts in Seoul showing how different vendors make use of the same unit, *c.* 2014–16. Pencil (2B and red) on A5 Bristol board. © Ray Lucas.

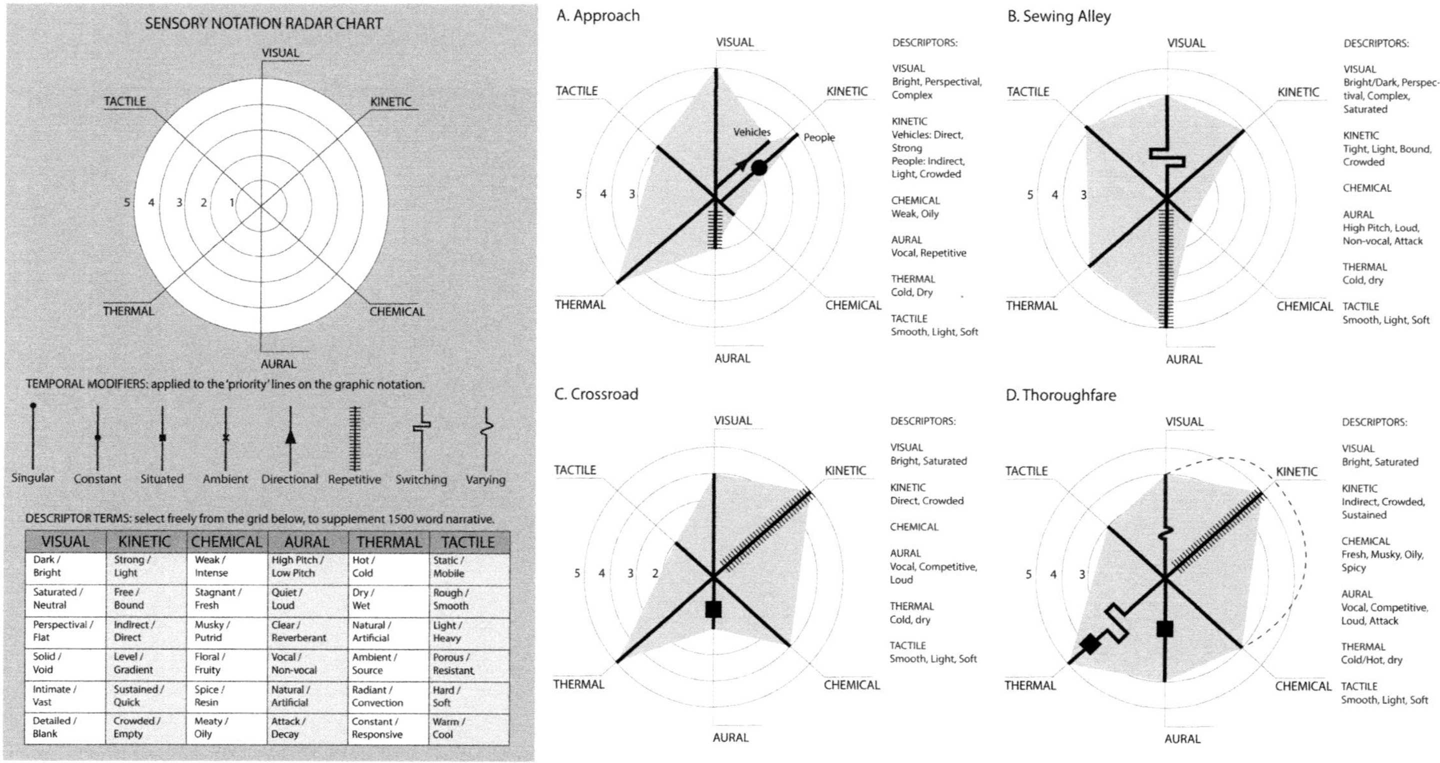

VISUAL	KINETIC	CHEMICAL	AURAL	THERMAL	TACTILE
Dark / Bright	Strong / Light	Weak / Intense	High Pitch / Low Pitch	Hot / Cold	Static / Mobile
Saturated / Neutral	Free / Bound	Stagnant / Fresh	Quiet / Loud	Dry / Wet	Rough / Smooth
Perspectival / Flat	Indirect / Direct	Musky / Putrid	Clear / Reverberant	Natural / Artificial	Light / Heavy
Solid / Void	Level / Gradient	Floral / Fruity	Vocal / Non-vocal	Ambient / Source	Porous / Resistant
Intimate / Vast	Sustained / Quick	Spice / Resin	Natural / Artificial	Radiant / Convection	Hard / Soft
Detailed / Blank	Crowded / Empty	Meaty / Oily	Attack / Decay	Constant / Responsive	Warm / Cool

Figure 9.4 Sensory notations of: A. Approach; B. Sewing Alley; C. Crossroad; and D. Thoroughfare. © Ray Lucas.

that these senses are either neglected in the design process or relegated to second-order representations. The following passage uses a combined strategy for notating the entire sensory experience of a space expressed. Rather than conforming to the classical notion of the five senses, these are combined somewhat differently as the visual, aural, chemical, kinetic, tactile and thermal systems. The language of perceptual systems draws directly from environmental psychologist James Gibson's ([1979] 1986) work on perception as an active and engaged process rather than a passive one. This section asks a question of how the market's surfaces present outside of their visual and geometric manifestations (Figure 9.4).

Visiting on a typical weekend market day near to noon in January 2014, the market presents the visitor with a combination of sensations.[10] The entry to the market from Myeongdong takes us along a stretch of roadway and pavement given over to pedestrians. The right-hand side is dominated first by the upmarket Shinsegae department store with its sober colours and careful design; and secondly by one of the large market complexes selling clothing. Along the left, a collection of slim, tall buildings with bright signage competes with a row of informal market stalls selling clothing. These are wide-based wheeled structures designed to have a broad surface area when fully unfurled to display goods. Often these are stacked high as a display of abundance, meaning that shoppers rummage and engage directly with the vendors to find what they are looking for.

A right turn reveals a side-street devoted to clothing alterations: rows of clattering sewing machines sit on benches overflowing with zips and fabric swatches, and gently swaying examples hang from metal frames above. The space is dark, punctuated by spot lighting, fierce blue-tinged LED lamps overhead angled individually by each seamstress. This space is crammed tight, with just enough room for customers to pass one another, but certainly no more than that. The proximity raises the heat a little, but the stall holders are working in thick puffer jackets swishing in concert with their swift, well-practiced movements. I complete a circuit of the alley, returning to the main market street at a crossroads.

The crossroads sits near the crest of a hill, and the orderly row of stalls has dissolved into a more chaotic jumble of stalls vying for the best position. The prospect shows the scale of the market as it drops off in the direction of the Sungnyemun city gate. The cross-street is occupied largely by informal stalls, arrangements of self-made frames and timber elements mounted on castors or large spoked wheels, making use of bright multicoloured parasols both for covering and for territorial definition, but also as additional hanging space. This places goods at various eye levels simultaneously, meaning that wherever one looks, there are goods for sale. On the fringes of the cluster of stalls are several compact food stalls selling a variety of local snacks, such as cinnamon-filled rice crackers, glutinous rice balls in a fiery chilli sauce and the pungent aroma of silkworm larvae.

The next section of the market is packed with modular steel carts provided by the city administration. These are stacked in a variety of ways, making use of the platform to display a wide variety of goods including jewellery, underwear, fried food, sunglasses, jeans, sportswear, shoes and toys. The street is packed tightly with shoppers, a channel down the centre accommodating two rows of stalls with a narrow middle section for vendors. The market buildings open out to the street with retail units, most of which are independent concerns – but some chains are joining this, often playing loud K-pop music and with PA systems announcing their cosmetics for sale. More typical are the traditional medicine shops with amber-coloured phials filled with ginger, ginseng and other roots. These also spill out onto the street, to a carefully managed extent. Shopkeepers maintain a presence on the street, often collaborating with neighbours to monitor their wares. The experience of this main thoroughfare is varied, bright and colourful, noisy and scented, with wares available to be touched.

Components and form in parallel projection

These unscaled drawings are detailed representations of the geometry and materials of a series of market carts and stalls (Figure 9.5). The convention of axonometric, part of the broader family of parallel projections, allows for the representation of a form in three dimensions with controlled deformation of angles. Where a perspective drawing has very few geometrically true angles and dimensions, all the dimensions of an axonometric drawing are true and in proportion with one another. The horizontal faces also retain true angles, meaning that a 90-degree corner is drawn square. The vertical faces are sacrificed, however, and these are deformed such that the vertical uprights are offset, sometimes giving a curious foreshortening to the drawing. The result is a drawing that has an unsettling aspect to it, almost model-like, giving the illusion of three dimensions without reproducing a view that one could have in reality. The series of drawings decontextualizes the stalls, setting them apart from each other and from their site. This is rectified by considering the serial nature of the drawings. Much like the earlier account of Cullen's *serial perspectives* relies on the sequentiality of the drawings to make sense, these drawings are also, as is commonplace in architectural drawing, to be understood as a *set*.

The drawings were first produced as a series of coloured sketches, testing the procedure for the drawings and establishing a protocol for their production.

Figure 9.5 Axonometric studies of market stalls from the series 'A Graphic Anthropology of Namdaemun Market', *c.* 2014–16. Pencil, pigment marker and fine-liner in A5 Leuchtterum Sketchbook. © Ray Lucas.

Following this through into the more-finished drawings with an eye towards their exhibition and scanning: drawing in a book makes their utility limited, so the move to specially prepared Bristol board was made.

Bristol board is a heavyweight machine-made paper used for illustration, originally produced in the early nineteenth century.[11] It is one of a family of illustration boards produced by laminating untreated papers together for stiffness and resistance to buckling. It is characterized by a smooth, white, relatively non-porous surface allowing media such as marker pens to be used without bleeding. The board has a weight of 250 grams per square metre (gsm) or higher, compared with typical cartridge paper which is 125 gsm. Like other forms of industrially manufactured paper, it can be produced in a range of weights and surface finishes. Bristol board is usable with a wide range of media, from pencil and ink through to watercolour and gouache (although traditional techniques for producing washes might misbehave due to the resistance and lack of porosity of the board: liquid media will pool on the surface rather than being absorbed). Its stiffness also makes it useful for collage and some model-making. It is a forgiving material but lacks much of the character of paper and board with more of a tooth to it: its smoothness makes it appropriate for some uses, but less so for others.

As I am working with architectural drawing conventions, it is useful to have reference points on the sheets of paper. When working with Bristol board, I prepare a grid of points in very pale grey on my computer, using an online tool which gives a range of options.[12] The points are 5 millimetres apart, and give useful guidelines for working to. Unless the viewer is looking for them, they recede into the background: particularly if a colourful marker sketch is completed on the page. I draw free-hand, but with reference to these guidelines.

Using a mechanical pencil with 0.7 millimetre lead, I begin to draw. At this point I am referring to a series of photographs from my field study. I normally use a soft lead of 2B or darker, as a harder graphite has a tendency to inscribe the marks into the surface more deeply, making errors more difficult to erase. The pencil drawing is the time for mistakes and discoveries to be made. Broad massing is the first task: having established the rough scale of the drawing, the proportions are sketched out by selecting the most significant forms. This process is one of establishing an underlying geometry.

This often reduces the scene to a series of boxes, which are then fleshed out with further detail. A rectilinear shape might have rounded corners added, or define the limits of a drum, but the broad composition is drawn first as boxes. Once this is in place, the process is one of gradually adding details and erasing 'hidden lines'. This is where the robust nature of the Bristol board is essential, as it allows multiple erasures. The forms are initially depicted as a kind of transparent X-ray, but as the drawing is established, lines which would not be visible from the presumed viewpoint are erased – or decreased in prominence in the next step of inking.

Inking is the stage of committing some lines to permanence and is rather more carefully executed because of this. For this stage, a variety of fine-line pens with a range of thicknesses from 0.1 millimetre to 1 millimetre are used. The thickness of the line can be used to denote materials being depicted – for example, steel poles are given a harder outline edge than a piece of cloth hanging down. I also use different colours of fine-line work at this stage, giving more options to communicate the qualities of the piece. The smooth drawing surface means that the lines are drawn precisely as intended. In a sense, Bristol board, fine-liners and markers are self-effacing materials: materials which do not seek to announce their presence in the way that charcoal or impasto oil paints do.

The final step is colouring. Using pigment ink markers, the resistant and non-porous nature of the board is again essential. This prevents the ink from bleeding, meaning that the coloured areas are accurate and intentional – with colours true to those of the ink. Using markers with more absorbent paper leads to a number of unintended consequences which might be interesting in a watercolour, but much more problematic in this case.

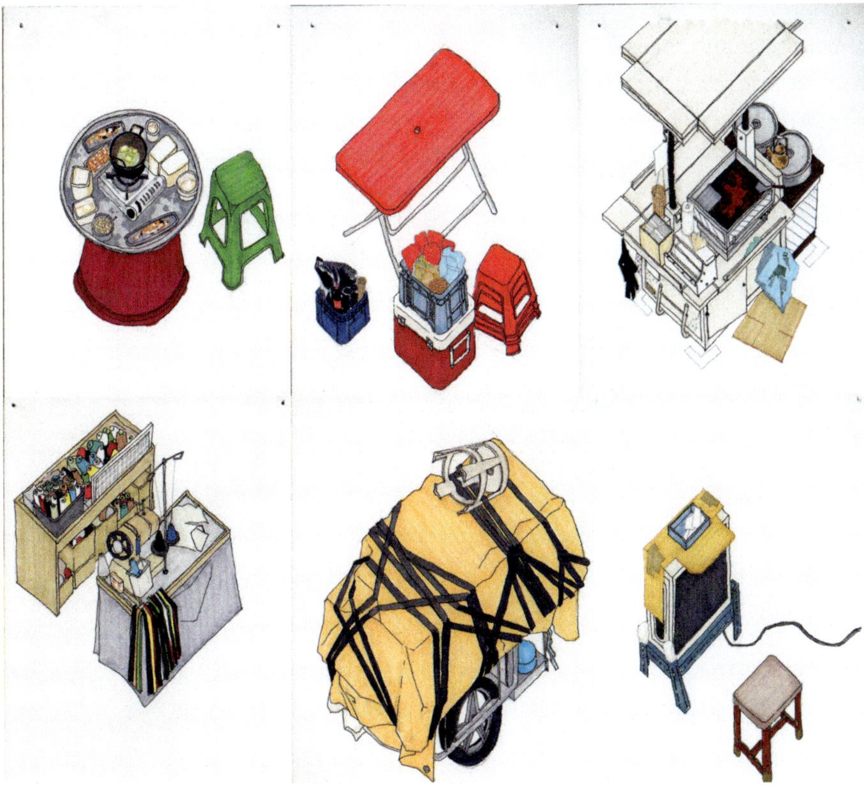

Figure 9.6 Axonometric drawings of market stalls from the series 'A Graphic Anthropology of Namdaemun Market', *c.* 2014–16. Pencil, pigment marker and fine-liner on A5 Bristol board. © Ray Lucas.

The robust nature of the board has a further set of affordances: the drawings can be exhibited as they are, without backing; can be copied and reproduced accurately; and can be handled without too many worries about damaging the work. This was another factor in the choice of Bristol board for these series of drawings (Figure 9.6).

Conclusion: Rehabilitating the superficial

Several versions of surface are discussed in this chapter, with the aim of rehabilitating the idea of the *superficial* in architecture. The marketplace is a site with a proliferation of surfaces at a variety of scales. Architectural drawing and notational conventions can be used to describe a broad range of surface conditions: from the literal material form-making conditions of architecture; to the borders between legitimate and illegitimate space; the multi-sensory effects of surfaces; and the means of passing through them via thresholds. Superficial architecture works with Gibson's use of the term: a superficial architecture is one of mediation.

Architectural drawing is often discussed in terms of its *generative* nature, as distinct from the representational or performative nature of fine art practices. This tells part of the story, and raises a question about the nature of mimetic, survey and copied drawings by architects: is this also a generative practice? I would argue that this is the case, and that the generation of architectural knowledge is furthered by the practice of drawing as well as the process of design. In the case presented, drawing is a practice of perception – a way of interrogating and understanding a place in order to find out more about it, to critique it and place it within the orbit of other, more conventionally understood architectures.

Surveys are often forgotten in the discussion of architectural drawing practices, in favour of the more explicitly generative drawings of the design process. I argue that these drawings, framed as graphic anthropologies, are generative in different ways: generating architectural theory and understanding rather than design. Architecture is both a design practice and an academic discipline, and this is one of its great strengths. Using the means of architecture to develop its theoretical side is one way of exploiting the distinctiveness of the discipline, increasing our value to others in cross-disciplinary exchanges as well as making theory *for* and *by way of* architectural practices.

Notes

1 I have written elsewhere about other aspects of this market: in an entry for *Our Lives with Electric Things* (Lucas 2017a), I discuss the use of electrically powered carts provided by local government as an instrument to limit the extent of the market by increasing reliance on infrastructure; in a contribution to an artist's book (2017b, 2017c), I turn my attention to the importance of some of the key materials used in the market, focusing on tarpaulin and castors. The drawings forming my ongoing drawing project of 'A Graphic Anthropology of Namdaemun Market' are presented in progress in *Drawing Parallels* (2019: 143–52), showing the frustrations and difficulties of the inscriptive practice and also used to discuss the gestural nature of drawing (Lucas, 2016). More substantially, the architectural anthropology of sites of exchange draws substantially on my Namdaemun fieldwork in *Anthropology for Architects* (2020a: 101–35). A natural partner for this chapter is in *Surfaces* (Lucas 2020b), where I discuss the proliferation of surfaces in the marketplace, and a brief entry in the *Encyclopaedia of the Vernacular Architecture of the World* (2021) positioning the market stalls as a form of vernacular architecture.

2 The work presented here leans heavily on architectural drawing conventions; particularly parallel projection. These rules allow a common understanding to be held between similarly trained individuals about what each line of a drawing means. The selected drawings in this case do not show human figures explicitly, instead relying on the spectator to fill in some of the blanks. This becomes an act of imagination, projecting the viewer into the field of an axonometric drawing, where other conventions will implicate the 'anthropos' more directly. I discuss issues of mimesis in drawing, and the uses of drawing by anthropologists such as Taussig (2011), Causey (2017) and Schwanhäusser (2016), further in *Drawing Parallels* (Lucas 2019a: 10–12) and *Anthropology for Architects* (Lucas 2020a: 25–44). Further graphic anthropology projects by the author have made greater use of conventions that describe the human body in great detail, such as Laban movement notations describing festival participants in Tokyo (Lucas 2018: 81–96).

3 A distinctly architectural approach to cataloguing similar spaces is taken by Bruno, Carena and Kim (2013); a variety of formal and informal market spaces are drawn, described and borrowed for their Venice Architecture Biennale exhibit.

4 Similar issues are raised with regard to another drawing support: tracing paper (Lucas 2017e), notable for allowing discussion of how the support can have a direct impact upon a drawing practice.

5 See Cullen (1961) and de Wolfe (2013) for examples of this method. Cullen was particularly vocal through his writings in the *Architectural Review.* His work belongs to a context of experimentation in urban design representation, including the *Imageability* studies of Kevin Lynch (1964) and Lawrence Halprin's *Freeways* (1966), and later *RSVP Cycles* (1982). Others involved in this broad effort at notational diversity in architecture include Philip Thiel (1996) and Mario Gandelsonas (1999), forming the basis of works by Tschumi (1994, 2014) and Allen (2008).

6 A worldwide movement engaged with in-situ drawing in urban environments. See Ridyard (2015) and other handbooks such as those by Bower (2016) and Campanario (2012).

7 Detailed by MacArthur (2007), the concept of the picturesque is the site of something of a culture war in architecture. This is promoted by a broadly conservative impulse for retention and historical context. Underlying this is a desire for continuity – the celebration of visual variety present in organic city planning as set against modernist tabula rasa conditions and efficiency of construction. This impulse belonged to a perceived need for rapid rebuilding of cities damaged by war along with a narrative of progress and modernization as a social good. The realities of embedded power structures represented by historical architecture, and the needs of contemporary cities, make for a much more nuanced and complicated story beyond simplistic ideas of conservation equating with the right-wing and modernism representing the left. Articles by the architectural historian Niklaus Pevsner, contemporary of and supportive of Cullen's work, are collected in *Visual Planning and the Picturesque* (2010), again highlighting visual methods of appraisal.

8 The nature of the variations possible in making these cuts is discussed by Lewis, Tsurumaki and Lewis (2016). The section is discussed in terms of its performance here, with the implications of actions such as *extrusion* and *nesting* for the design process. Leonidas Koutsoumpos (2018) notes that the section drawing as slice or cut emerges at a similar time to dissection practices in understanding human anatomy.

9 One of the few convincing histories of this tacit architectural knowledge is given by Guillerme, Vérin and Sartarelli (1989), arguing for a connection with the fascination for ruins and tracing the history of this drawing convention, which effectively slices through in order to reveal the interior of a building; the rules of the representation are similar to elevations, where frontality and flatness allow angles to remain true and measurable, without any of the deformations present in perspective or parallel projections. The effect is a flattening of space, sacrificing the illusion of three dimensions. Others such as Robin Evans (1996) have explored short-lived conventions (also named 'sections') which serve a similar purpose of showing interior spaces, but through a folded-down internal elevation arranged around a regular plan drawing.

10 This section uses the *Sensory Notation* system developed by the author (Lucas and Romice 2008, 2010; Lucas 2009a, 2009b, 2009c). Simple radar chart diagrams are used to describe sensations in terms of how strong they are, rather than to indicate positive or negative aspects of the experience: a pleasant scent can, for example, be ranked as highly as a deeply unpleasant one. The graphic is supplemented by a plan and section, a textual description and photographs of the triggers of sources for sensation. The notation includes corroborations and temporal indicators shown in the key as well as a series of 'descriptor terms' designed as plain and straightforward shorthand to the conditions. The aim is to replicate the experience of using a sketchbook to record the visual aspects of a site, so there is room for imprecision and variation between notators; the notations can be used either as precedents to be replicated elsewhere, to diagnose problems with the sensory qualities of a space or to find out how multiple participants experience a site simultaneously.

11 This section on Bristol board was previously published in Lucas (2017d).

12 The source for the grid paper tool is: http://incompetech.com/graphpaper/squaredots/ (accessed 9 August 2019).

List of Figures

Chapter 10
Archive Surface

Jane Birkin

Archive objects are fascinating from the perspective of their diverse surfaces, as well as the surfaces of the envelopes, wrappers, files and boxes that contain them. These archival interfaces increase our sensory awareness of the diverse objects held, and mark the physical archive as a place of textural as well as textual information. They also allow consideration of objects outside of the standard information and evidence-based archival structure, an epistemic framework that Lisa Gitelman (2014: 1), in her media history of the document, terms 'the know-show function', thus defining the purpose of the document as being *to* document.

A recent project provides me with the opportunity to discuss a particular archive object, one that is of special interest because of its materiality and its uniqueness of surface. Although the object in question – a bundle of letters – is made of paper and should be part of what we class as the 'paper trail', it is so damaged that the written evidence that it contains is inaccessible. Yet this object opens up a complex dialogue around surfaces, in terms not only of their physical make-up, but also of how we think of them in relation to the complex notion of 'document' – how we navigate the tensions between materiality and information. The project that forms the backdrop to this discussion is one to create a 3D surrogate of the object, a non-interventional scanning and replication process that allows a documenting – and a preserving of sorts – of the outer surface of this object. More advanced technologies – such as those involving X-ray tomography, developed by the École polytechnique fédérale de Lausanne and used with objects from the State Archives of Venice, for example – may be able to reveal what is inside sealed and degraded objects, but these technologies are not widely available to archives, and nor are they appropriate to all objects. Moreover, the point of this project is to better understand the object through its surface, using technologies and expertise in the field of 3D reproduction as developed within an art and design context.

New surfaces are created and met in the course of the 3D scanning and printing processes: the digital image of the object turning and floating on the surfaces of different screens, and the 3D print itself, first as soft, raw gypsum and then augmented with a stronger surface of glue. Most importantly though, this technical process and the resulting dialogue between original and surrogate offer a fresh way of looking at this unique archive object – at what its surface reveals as well as what it conceals. Although there are no immediate plans to repair this particular bundle, and it is considered here in terms of its current state of decay, I will nevertheless go on to briefly describe how this might be done, from the perspective of the many different surfaces that are encountered through the preservation and conservation procedures of objects like this one: the original surface of the bundle, the discrete surface of each letter and the new surfaces applied.

But before we even arrive at the object itself, many surfaces must be negotiated: the archive is an overtly corporeal and materially controlled information management system, and is dependent on the maintenance of surfaces to uphold its power over the documents it houses. The walls of the archive strongroom form an outer membrane, a protective surface that fully encloses the archive materials, keeping them away from the touch and the gaze of the public. This internal zone is a private space of historical arrangement – objects within it are stored in stasis, frozen in time through complex archival cataloguing processes and concordant location guides. In contrast to this arrangement, when materials are requested and fetched out for public view they enter the ever-contemporary milieu of the reading room, where they may be reconfigured and realigned, depending on the focus of the reader. The administrative apparatus of the archive ensures that after use these objects return quickly and safely to their assigned place inside the file, inside the box, inside the strongroom. Sarah Oppenheimer (2014), in her interview with Giuliana Bruno, sees surface as 'the membrane that subsumes the divisions of interior and exterior, past and present, public and private'. These are the divisions – the institutional surface tensions – that sit at the heart of the archive.

The Wellington 'bundle' and cultural 3D

In common with many institutional repositories, the University of Southampton Library's Special Collections, a place where I spend half of my working life, holds various kinds of objects that run across diverse ranges of archives and collections. For the most part, these objects – papers, photographs, rare books, textiles and more – are in good condition. They are catalogued and described, and storage methods and environmental conditions ensure that they are well preserved. In addition, many objects have been painstakingly cleaned and repaired in order to enable their safe and continued use by researchers. There exists within the Archives and Manuscripts division a number of tightly folded and tied bundles of letters. These objects are of special interest within a dialogue on surface – and on the related notion of superficiality, a trait not usually associated with archival and institutional thinking. The bundles, collected together by year and month, were included in the papers of Arthur Wellesley, the first Duke of Wellington (1769–1852), when they came to Southampton in 1983. The Wellington archive is vast, containing around 100,000 items, covering the period from 1790 to 1852, and including political, military, diplomatic and official papers of great historical significance. Wellington was one of the leading military and political figures of the nineteenth century, but it is probably his role in the Peninsular War, and especially his decisive victory over Napoleon at the Battle of Waterloo in 1815, for which he is best known. The papers saw challenging times before they arrived in Southampton: some were damaged in a shipwreck in the river Tagus in 1814, as the ship bringing them back to the UK from the Peninsular War sank as it left Lisbon; others were affected by damp while in temporary storage during the Second World War.

Although the letters from many of these bundles have been extracted, cleaned and flattened, some remain in bundle-form. Many of these are in relatively good condition and can be safely unfolded and read, but extensive mould damage has left several bundles of letters fused together, fragmented and crumbling. I am going to focus attention on one of these bundles (Figure 10.1): one that is particularly degraded, to the extent that it can be handled only by conservators – and even this is avoided. This bundle certainly cannot be unfolded, and therefore the letters that make it up cannot be read. Its current condition is preserved by way of modern archival storage techniques: it is wrapped, boxed and kept safe

Figure 10.1 Wellington bundle, *c.* 1832. Wellington papers. © University of Southampton, Special Collections.

in the environmentally monitored and controlled atmosphere of the strongroom. When the tissue wrapper that envelops the bundle is folded down, all that can be seen is its outer surface: the relatively stable top, and the dusty and decaying edges of the folded letters at the sides and ends, where a few fragmented words are visible. The underside cannot be viewed at all, as the bundle itself cannot be lifted up. As well as having been poorly stored in the past, this archive has a history of disastrous rearrangement, sporadic documentation and mislaid indexes, with only brief catalogue descriptions of the letters surviving, so all we know about the bundle is that it contains letters to and from the Duke, labelled April 1832. At the time that these letters were written, Wellington was Colonel-in-Chief of the Rifle Brigade, Lord Lieutenant of Hampshire, Constable of the Tower of London and Lord Warden of the Cinque Ports. His two periods as prime minister, from 1828 to 1830 and then again briefly in 1834, bracket their writing.

The 3D-printing project that took place around this object was a joint one with the Winchester School of Art, specifically with artist Ian Dawson and technician Chris Carter, who both have excellent knowledge and understanding of 3D processes, especially with regard to the 3D printing of historical and cultural objects. Dawson's interest lies in the materiality of the medium of 3D print, including through experimental scanning of broken spaces and surfaces, as in his Taplow House project (with Louisa Minkin), where interiors and discarded objects from a derelict south London housing estate were scanned and printed prior to demolition. Dawson has worked on other projects, such as one that culminated in the exhibition *The Wanderer's Nightsong II* (2015), where he acted as artist, curator and technician, 'translating the work of each artist via 3d scanning and printing into copies of themselves – asking questions of mediation, re-mediation, authorship and originality' (Dawson 2015). He has also worked in collaboration with archaeologists, scanning artefacts and creating images, considering the original object through an opening up of its surrogate digital surface.

This 'bundle' project is one that combines the artist's engagement with the 3D medium and the archive's focus on the document – *and* the documentation of the document – always with an interest in developing new documentary processes. In light of its complex and currently impenetrable exterior, all parties were agreed on the suitability of this object for investigation. As an artist working within an archive, I was able to consider the value of this project from both sides. When three of the damaged bundles were brought from the strongroom and put under studio lights for capture, they appeared strange and stirring to those who had not seen them before. For those of us who were familiar with these objects, the project allowed them to be seen in a different context altogether. Until this point, they were considered for the most part as degraded objects that were interesting both for their historical significance and also for the challenges they presented in terms of conservation. It quickly became apparent, to me at least, that they also had much to offer in terms of exploration of the irregularities, inconsistencies and contradictions of their physical surfaces. As Tim Ingold (2017: 100), in his essay 'Surface Vision', states in his plea for our attention to surfaces: 'There is so much to see; so much to learn!'

The method of capture used for this project was photogrammetry – a topographic mapping system whose origins lie in the nineteenth century, not so long after the writing of these letters. It is now widely used in conjunction with drones for agricultural monitoring, as well as by the military. The 1832 bundle was photographed from many angles, and the resulting images were put through a software program to be combined, in order to provide an accurate measurement of surface points (Figure 10.2). The new computer-generated object could be turned and flipped, the zero-thickness form of its digital surface transforming it into a lithe and supple representation of the original. In real

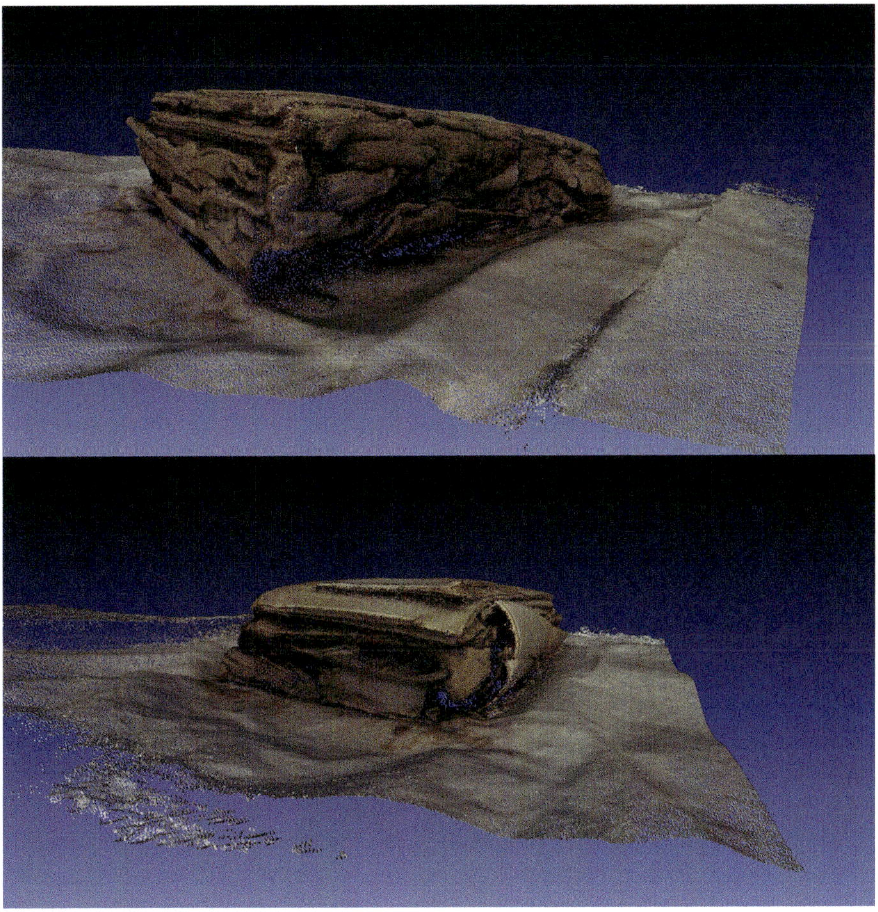

Figure 10.2 Ian Dawson, Wellington bundle (scan), 2017. © Ian Dawson.

terms, the scan itself is simply a data set, as Hito Steyerl (2017: 192) explains in 'Ripping Reality: Blind Spots and Wrecked Data in 3D': '3D scanners generate point clouds, measurements in virtual space that can in turn be rendered as 3D objects and printed.' Jacob Gaboury (2015: 45) explains that images are materially structured by the logic of computation, relying on 'visual tropes such as perspective to structure its analysis, a structure that obfuscates much of the material workings of computer-generated images'. And Vilém Flusser highlights our capacity to *read* the surface of an image:

> Images are significant surfaces. Images signify – mainly – something 'out there' in space and time that they have to make comprehensible to us as abstractions (as reductions of the four dimensions of space and time to the two surface dimensions). This specific ability to abstract surfaces out of space and time and to project them back into space and time is what is known as 'imagination'. It is the precondition for the production and decoding of images.
> (Flusser 2000: 8)

All this activity, the underlying computational processes of measuring, calculating and perspectival rendering – and the human 'decoding' of the movement on the surface of the computer screen – formed a complex period of production, with bundle and image demanding equal attention and consideration as cultural objects.

Paper as surface and interface

In a 'real-life situation' – in contrast to a replicated one – Steyerl (2017: 200) argues that 'the surface will bear the imprint of the political, material, social, technological and affective forces that shape it'. In support of this argument she cites Siegfried Kracauer, who perceives the surface as a primary site of historical and social information, and who breaks with 'the more traditional view of surfaces that associates them with superficiality' (ibid.). This bundle has been shaped by all of the forces that Steyerl lists. Historical evidence through *reading* gives way to the evidence of the passage of time since *writing* and presents an understanding – and a documentation – of time in an overtly material way. In terms of pure information, this object could be viewed as superficial, even non-existent. Of course, there is no such thing as pure information, as is now widely accepted in terms of the digital, although Gitelman, citing Geoffrey Nunberg, argues that 'information has an objective, an autonomous character, partly because of the way it reflects the authoritative institutions and practices to which documents belong' (Gitelman 2014: 4). And the way that we *process* information, in small chunks, reflects the way that it was primarily (before the 'information age') brought to us in a discrete way, on pieces of paper (ibid.). Here, in the absence – or inaccessibility – of information of the institutional variety, the surface, the decaying paper support, becomes the dominant feature. In *Paper Machine*, Derrida (2005: 41, original emphasis) argues that he has never had any other subject than paper: 'I have always written, and even spoken, *on* paper: on the subject of paper, an actual paper, and with paper in mind. Support, subject, surface, mark, trace, written mark, inscription, fold – these were also themes that gripped me with tenacious certainty.' But in a typically contradictory fashion, he challenges the prevailing view of paper as support:

> The word *support* could give rise to plenty of questions on the subject of paper. There is no need to trust blindly in all the discourses that reduce paper to the function or *topos* of an inert surface laid out *beneath* some markings, a substratum meant for containing them, for ensuring their survival or subsistence. On this commonsense view, paper would be a body-subject or body-substance, an immobile or impassable surface underlying the traces that may come along and affect it from the outside, superficially, as events, or accidents, or qualities.
>
> (Ibid.: 42, original emphases)

Derrida (ibid.: 44) is framing paper as multimedia, the page as screen, with information contained under and in the surface: 'beneath the appearance of a surface, it holds in reserve a volume, folds, a labyrinth'. Paper is undeniably a multimedia material: a transmitter of information that reaches far beyond the idea of the writerly text (Barthes 1990: 4) in terms of interactivity, and into the broad spaces of the sensorium.

When Derrida (2005: 44, original emphasis) wrote *Paper Machine* in 2001, he perceived paper as on the wane: 'Paper is *declining*, it is getting smaller, it is shrinking inexorably at the rate that a man grows old.' Today, we are generally more optimistic. In the post-digital era, we have developed something of a fresh appreciation of material objects, and the archive plays a part in this – not the objects alone, but the material and labour-intensive processes and systems that support them. Bruno (2016) explains:

> It is interesting to note that an intense interest in material culture and a deep fascination for the materiality of the archive has emerged in the digital age, characterized by networks and their seeming immateriality. New disciplines, methods, and forms of scholarship have originated from this tension between material and immaterial culture, and they express this very tension in their development.

She links this to media archaeology, as it presents 'deep interest in the material history of things combined with a fascination for systems and networks' (ibid.). Thus, media archaeology's link with archives is well defined.

Alessandro Ludovico (2012: 7) argues that digital is built for speed, and print for stability. And if you are in any doubt that paper is not stable (and you might be, given the state of the bundle being discussed here), or that digitized archives are effective backups for physical ones, then consider how the digital file is at risk, both inside and outside the archive: the decay in software/hardware systems is material, latent and inescapable. One must also reflect upon the conceptual implications of digitization for the archive, as the prevention of digital collapse requires the updating of software and hardware systems, a rewriting that is fundamentally alien to the archivist's desire for permanence of storage and materials. Wolfgang Ernst (2015) identifies this as 'a change from the ideal of archival eternity to permanent change'. Paper, on the other hand, especially old paper – as long as it is not subject to shipwreck and the like – is a remarkably stable medium; parchment even more so. These are the materials on which archives – and their systems – are built.

Ludovico (2012: 7) puts forward a reason for the survival of paper in the context of the plethora of digital platforms and devices, not in terms of its material stability, but because of its popularity:

> And yet, there are still plenty of newsstands and bookshops around, well-stocked with a wide variety of printed products. And if you are reading these words on paper (which you probably are) then you have, for some reason, chosen to go with the 'old' medium. Why? Probably because it still comes with the very best 'interface' ever designed.

The words 'surface' and 'interface' are used simultaneously within fields such as life sciences and chemistry, including molecular and atomic science (Allara 2015: 638–9), as well as in computing, where the surface/interface of the computer screen now almost entirely facilitates our engagement with computers. This is what Gaboury (2015: 57) terms 'the computer's disavowal of its own materiality through the black boxing effect of the interface'. He makes a plea for us to look under the surface of computer graphics, not only to the code, but to the materiality of the visual objects and their perspectival and illusional support systems, which he terms the 'hidden surface problem'.

Icons of the archive: the dust and the secrets

Media theorist Cornelia Vismann (2008) engages with problems of what the surface may hide as she recounts how an archive dating from the third and fourth centuries was discovered in Greece in the 1970s. This find consisted of several lead tablets, each detailing on the outside a man's name, and a description of his horse and its value. The intriguing thing about these tablets is that they appeared to never have been opened, even at the time of their use. So of what use were they, and what use are they now? Vismann draws a comparison between these finds from antiquity and German artist Anselm Kiefer's lead books series, such as those in his work *The High Priestess* (1985–9), which are so heavy that they cannot be opened or even moved. Vismann (2008: 162) writes of the Greek tablets: 'They are files at a standstill ... they are an aesthetic monument. For what is one to do with these unreadable tomes other than venerate them as icons of writing and literacy?' This archival bundle might similarly be perceived as an aesthetic monument, as an icon of literacy – iconic because it is unreadable: we are de facto drawn to its surface materiality.

In *What is Media Archaeology?* Jussi Parikka (2012: 117) argues that everything decays, and that material decay is a sign of radical temporality, and therefore a topic of media-archaeological research. Alongside such critical research, degradation and decay must also be considered against the background of the visual culture and the popular perceptions of the archive. Decay is expected of archive objects; they are not considered spoilt by decay, but somehow enhanced. This object might even be considered an icon of the archive, fitting within an aesthetic of surface dust and material decay. Parikka (2015: 85) also writes about the aesthetic and even the romantic qualities of dust:

> There is something poetic about dust. It is the stuff of fairy tales, stories of deserted places – of attics and dunes, of places from so long ago they seem to never have existed. Dusty books: the time of the archive that layers slowly on shelves and manuscripts. Marcel Duchamp's 1920s Large Glass was a compilation of dust. He allowed dust to do the work.

The dusty aesthetic of the archive, which extends beyond the objects themselves and into the corridors and storage systems in which they are contained, is by and large, as Parikka acknowledges, an imagined one: with the use of hi-tech environmental management systems and HEPA filters, surface dust in the modern archive environment is something of a myth. However, there is no contesting that this particular object conforms to popular expectations of an archive object: it is old, it is dusty and it is degraded. Like Duchamp's *Large Glass*, it is in part made up of dust: dust does not just lie *on* its surface, it *is* its surface – and dust is measured and recorded in the 3D scan as systematically as any more stable part.

In spite of its obvious richness – its temporal and aesthetic attributes – if one is to view the archive purely as a site of information, of knowledge formation within the institutional framework, this bundle clearly does not perform according to expectations. As Gitelman (2014: 5) argues: 'Documents belong to that ubiquitous subcategory of texts that embraces the subjects and instruments of bureaucracy or of systematic knowledge generally.' In this context, then, our bundle is an overtly superficial document; at best a possibility, at worst a mere representation of information. Moreover, superficial or otherwise, the fact is that the object remains unavailable to researchers: they can neither read its inner contents nor admire its surface aesthetic qualities. In archives-speak it is 'unfit for production', a 'closed' object – and it is literally closed, as decay has penetrated to such an extent through the multiple surfaces of paper that they are fused together. What were once discrete and enumerable letters are now designated as a single archive object. Since there is no surviving catalogue record of these letters, they have also rather poetically become what the International Council on Archives define in their document *General International Standard Archival Description* (2000: 12) as the 'smallest intellectually indivisible archival unit' – and indeed what Foucault (2002: 90) in *The Archaeology of Knowledge* might term a statement, the 'ultimate, undecomposable element that here can be isolated. … The atom of discourse'.

Although the written information contained in the letters that make up the bundle might be inaccessible, it is not yet lost; it is simply inoperative, and therefore secret. Secrecy, like dust, forms part of the enduring view of the archive. It is popularly perceived as a closed and rather mysterious place, perhaps difficult to access but tantalizingly jam-packed with secrets waiting to be discovered. As Flusser (2011) tells us, the history of letters is anyway one of secrecy – and one of waiting, and these are, once again, letters in waiting. For Flusser (ibid.: 106), this secrecy is all about closed and inaccessible outer surfaces – the envelopes they are sealed in and the postboxes they are put in to – that relate to the cybernetic secrecy of 'black-boxing': 'Letters are sealed and thrown into black boxes (painted yellow, red, or blue), to be drawn out of black boxes somewhere else (e.g., post boxes) and opened. The whole process is secret, steeped in epistolary secrecy.' The outer surface of this bundle indeed acts like a corrupted file, an icon that is tantalizingly visible but refuses to be opened and read. Flusser (ibid.: 108) also reminds us that although letters are writings that are not intended to be widely accessible, they may nevertheless find their way out and into the hands of the public. One might ask, therefore, whether we should even be thinking of reading these letters at all. Indeed, at the time of their writing, these papers were considered to be the sole property of the state official to whom they were addressed. Envelopes and wrapping sheets, secured with seals, once formed outer surfaces that kept letters such as these secret, but the archival process opens them up to the public eye, as these letters, even though they might be materially inaccessible (like the letter in question here), are now deemed accessible by law. Chris Woolgar, head of Special Collections at Southampton when the papers arrived, explains that there has been a change in perspective, from a situation where the papers of state officials, diplomats and military personnel were considered their own private property, to the one that exists today:

> Today's expectation is very different: the records of government business belong to the Crown. They are accessible under freedom of information legislation and other avenues, and a selection is ultimately placed in the National Archives. This perspective is closely linked to a transformation in the way government was conducted in the twentieth century, from official business carried out in a private, or semi-private and informal way, to methods of working which guaranteed the secrecy of official papers on the one hand, and which have come to emphasise public accountability and scrutiny on the other.
>
> (Woolgar 2008: 4)

Likewise, Derrida discusses the implications of the archive in the 'politics of the secret', linked to state power and political exploitation. He argues that certain archives must not remain inaccessible, and sees this as becoming an even more burning issue with advances in police and military technology and 'the new problems of cryptography' (Derrida 2005: 162–3).

Complexion, complexity and affect

In its present state, this bundle is a contradictory object. Although its interior is a physically dense yet unknown entity, it is outwardly, materially, delicately complex. It possesses what Tim Ingold (2017) calls 'complexion', which he contrasts with simple 'surface'. He sees 'complexion' as a better word to use in relation to surface, rather than the word 'texture'. Complexion, as we know, usually refers to human skin, and more specifically to the skin on the face. As such, it is much more than a surface texture; it is intrinsically and intricately connected to the body's internal mechanisms, a visible indicator of what is going on inside, both physically and emotionally. Ingold (ibid.: 103–4) argues in relation to complexion: 'Here, interior and exterior are not set apart by the surface.' In science, surface is today considered an important field of investigation – not simply a top layer, the shell of something more important, but a significant, albeit super-thin, material in its own right. Magdalena Helmer (2005: 637) states: 'Although their significance has been realized for centuries, surfaces and interfaces long evaded detailed scrutiny at the atomic scale. After all, they are not simply the final layer of, say, a piece of metal or liquid in contact with air, but an exceedingly thin region with properties distinct from those of the bulk material on either side.' Ingold (2017: 103–4) argues along the same lines: 'The surface ... is the guarantor that inside and outside keep to their respective domains and do not mix. No surface; no object.' And Bruno likewise sees surface as 'a zone of encounter between us and the space that surrounds us' (Oppenheimer 2014). The notion of the existence of matter both inside and outside of the surface has implications for the archive object: the wrapping, the controlled environment and the clean hands (or the ubiquitous white gloves) that archives insist upon are as significant as our own skin in terms of protection against chemical reaction to light, temperature and airborne impurities. What is completely obvious in the scientific realm is that surface is not superficial in the commonly understood sense of the word, as in slight, insignificant, frivolous, cosmetic rather than complex, in short, not really worth considering at all. Ingold argues that it is this kind of thinking that is the reason for our distrust of surfaces, and why we feel we have to break through them to reveal anything of importance. Breaking down the association between surface and superficiality, and uncannily presaging what may one day transpire with our bundle-object, he asks: 'What if surfaces are the real sites for the generation of meaning? Then by mining them, excavating them, or clearing them away, we may in fact be destroying precisely what we seek to find, and that lies under our very noses, convinced as we are that the truth can never be on the surface but somewhere deeper down' (Ingold 2017: 100).

This bundle, this 'icon of writing and literacy', is not a conceptual object by intent like Kiefer's lead books; but it can be argued that it has been transformed by time and circumstances into one, into what might be broadly called an object of conceptual writing – if we agree with Kenneth Goldsmith's (2012, original emphases) statement that conceptual writing 'invokes a *thinkership* rather than a *readership*', that once the *system* is understood, the *words* do not matter. Citing Laura Marks's theory of 'haptic visuality' – a modality of vision that functions in similar ways to touch in invoking experience and memory – Ingold (2017: 101) questions the limits of hapticity: 'But is it limited to the hand? Can I not also wipe the surface with my eyes? Could eyesight be as haptic as manual touch?' This bundle is an object of such surface richness and depth that we crave to touch and feel it; but vision has to suffice.

Yet there *is* more: even from the image reproduced here (Figure 10.3), we begin to smell and even to taste its dust and desiccation. This muddling of the senses is confirmation of haptic visuality as an invocation of the memory of past experience, and equally of Goldsmith's notion of thinkership, of understanding the system – of understanding how and why an object exists within our amassed knowledge of similar objects.

This bundle can also be understood within the system of the archive itself. Goldsmith's (2011: 34) views centre on the duality of the text: how it flips back and forth between meaning and materiality, between formal qualities and communicative ones. These are problematic concepts in themselves, as materiality can carry meaning (temporal, historical and environmental), and the formal qualities of surface can communicate: they bring with them *affect*. As Bruno (2016) argues: 'As forms of materiality that touch us and can be touched, surfaces affect us. And it is in this surface intimacy that affects become revealed: surfaces affect us also because they retain the stains of time. Objects have their histories written into their surfaces.' Yet Goldsmith's (2011: 34) argument around

Figure 10.3 Wellington bundle, *c.* 1832. Wellington papers. © University of Southampton, Special Collections.

language – how 'we can choose to weigh it *and* we can choose to read it' – is key to the archive, where physical surfaces, recto and verso, folds, cockles, creases and tears afford a unique understanding of an object, one that is different from reading from a transcription, or even from a digital image. As Woolgar (2008: 33) remarks in the context of the push for digitization of archive holdings: 'In all of this, it is necessary to understand the way the documents themselves work, and what they can tell us from process and form as much as from content.' As a tangible footnote to all of this, at Southampton, documents are actually weighed as they are given out to readers, and then as they are taken back – a security measure that is only possible because of the materiality, the weight, of the varied surfaces of these archive objects.

Surfaces recreated

At present, however, Goldsmith's duality is absent from this particular object; it could be weighed, either physically or intellectually, but it is completely unreadable – and this brings us back to Vismann's question, 'what is one to do with these unreadable tomes?' When the papers came to Southampton under national heritage legislation, they came with the expectation that they would, where possible, be conserved, in compliance with the regulations on freedom of information that govern such material. And so the official 'what to do' with this object is to render it readable – the individual documents might be separated and the jigsaw puzzle of paper fragments put together. This has already been done with other damaged bundles, although our particular one is at the high end of the scale, which is partly why it remains in its present state. If work is ever done on this bundle – and it is expected that some of the more damaged bundles will be left as evidence of what has been – then there is a procedure. It is a painstaking and lengthy task, and much more complex than can be explained here, but it has been tried and tested over the years. The bundles are divided into smaller sections and left for a period of up to two years, in order for pressure caused by the load of their own compacted mass to alleviate. After this time, the individual letters are teased apart, where possible with a soft brush, or if not then by gentle humidification. Loose fragments are carefully realigned, and missing areas sympathetically filled in with a paper pulp similar to the original, using a mechanical technique called leafcasting. Finally, the whole surface of the letter is strengthened – or we might say that a new surface is formed – with a layer of gelatine or methylcellulose size. This replicates the original surface, where the paper would have been sized with gelatine.

When reconstructed, the letters become once again discrete documents that can be individually enumerated, described, transcribed and catalogued; the paper returns to its original role for Wellington and his epistolary associates, an everyday support for reading and writing. 'Support' is a term used meaningfully by conservators to describe the paper, parchment or other fabric part of an object – not only the surface of it, but its entire structure – as distinct from the ink or pencil used. Conservation treatments of the support medium and the graphic medium have to be considered differently, but in practice the two are often indivisible. The iron gall ink used on these and similar letters from this time is physically melded into the substance of the paper. We can see at magnification what Derrida (2005: 42) has already told us: the paper surface is neither 'inert' nor 'impassable'. The ink forms a mechanical bond with the paper, penetrating the spaces between the fibres and remaining firmly entangled in them when it dries; this is exactly what Derrida meant when he questioned the role of paper as a substratum for sustaining markings – for him, the two were as inseparable as body and soul (ibid.: 42–3). A soul of sorts, at least a system of memory, can also inhabit the non-human bodies of objects: as Bruno (2016) argues, 'there is also a body of things, even immaterial ones, which themselves have a vibrant material consistency and form a corpus of aggregations and systems of connections that are also mnemonic'.

Whatever the nature of the letters contained inside, the bundle itself is a unique object. With the separation and the reconstruction of letters there comes a feeling of satisfaction and accomplishment for the conservator, but also a feeling of loss for the object that was. Enshrined within the ethical canon of modern conservation is the idea that treatment and repairs should be reversible, and although only glues, papers and sizes appropriate to reversal are used in the rebuilding of letters – and so could in theory be removed – it is clear that an object such as this could never be reconstructed: the surface that is visible now would be gone forever. For this reason, and for reasons of archival and scientific rigour, the documentation of conservation projects – pre-, during and post-treatment – is routine and thorough, in the form of written reports and high-quality photographic evidence. And the photographic reproduction of *already* 'readable' surfaces is one way of circumventing the decay of the material object, preserving the information it is carrying, disseminating it, and protecting the object itself from the hazards of handling – although the materiality of the digitized object and its predisposition to decay are the source of great anxiety in archival circles, as has already been discussed.

To return to the 3D scanning and printing that initiated this discussion, the question arises: what if the purpose of digital reproduction in this case is not to enable the dissemination of written information, but to preserve – to record *for* the archive – the existing state of the decayed object through the creation of a 3D surface that replicates the original? Steyerl (2017: 191), with a line of questioning that implies the inherent superficiality of the 3D print, asks whether the traditional role of representation that comes with the photographic image is lost (or perhaps given more weight) in the process of replication by the creation of this simple 'stand-in'. Although questions of representation are not overly important to the archival condition, which places emphasis on objectivity and evidence, the *documentary* role of 3D *is* important, and Steyerl (ibid.: 192–3) asks: 'So what does the notion of documentary mean if applied to the 3D replication of objects and situations? What is the relation of 3D technologies to traditional ideas of documentary evidence? How are notions of documentary truth updated or displaced by 3D technologies? How does the ability to create 3D reproductions affect ideas about documentary truth?' One might expect that the addition of this third dimension would provide a more meaningful and informative document, and in some ways it does, as the bulk, the presence and the mode of viewing (we can move around it) are reproduced. Yet, 'the pleats and folds that constitute the fabric of the visual', to which Bruno (2014: 4) urges us to pay attention in our consideration of surface, are the very things that generate difficulties in the scanning process. The blind spots, caused by folds, shadows and angles of view, mean that information from the scan needs to be augmented and interpreted – either automatically by the software, or by the technician working with the software – in order to allow a viable print to be made, under the laws of physics. As Steyerl (2017: 197) explains: 'The missing data are assigned a volume or body. … What emerges is not the image of a body, but the body of an image that presents itself on a thin surface of differentiation, shaped by different natural, technological or political forces.' There is no need here for any engagement with the 'hacks, cracks and viruses for 3D print software' that Morehshin Allahyari and Daniel Rourke (2016: 329) advocate in 'The 3D Additivist Manifesto' in order 'to introduce errors, glitches and fissures into 3D prints'. Interestingly – and perhaps *superficially* – the fragmentation and surface glitches that occur in the scan due to missing data are analogous, but not exactly faithful, to the fragmentation of the object itself. The flaws in the capture of this difficult object – exacerbated by the technical limitations of printing – mean that this project should perhaps be classed, as Steyerl (2017: 193) classes her own (very different) experimental project, as 'a model to test these technologies for potential documentary practice'.

After the scanning and the computational rendering, the surface pattern of the object – the colouration and the markings visible in the original object, which are also the result of photographic capture and software consolidation – is at first kept separately from the point cloud measurements that generate the 3D form. This pattern is applied to the skeletal framework of the bundle. It is chosen from a bank of surface patterns that are stored on the computer from previous scanning projects – images of absolute superficiality. As the original bundle could not be lifted off its base, the underside was not scanned at all, meaning that the computer could neither read nor write the true volume of the object. The inner surface of the digital object is a fictional surface: it materializes only as a mirror version of its outer pattern. In order that the print can be made, the technician must build this inner surface, in what can only be described as the digital equivalent of throwing mud at a window. There is an intriguing link here with the reconstruction of the paper surface by the conservator – the leafcasting and the sizing – but then, as Ingold (2017: 104) argues, 'no surface; no object'.

The digital object is split into slices for printing: the more slices, the more accurate the print. These slices can even be likened to the contents of the bundle, that is, the closely compacted sheets of paper that form the solid object. Moreover, with each pass of the printer, the printed layers of gypsum begin to appear as an echo of the paper fragments that we see in the conservation studio. Again, this is a fascinatingly superficial manifestation, and one that is very particular to this object. The print itself is made of gypsum combined with a binder and with ink added from standard inkjet printer cartridges to reproduce the colour. Its surface is soft and fragile. In order to give it stability, it has to be 'cured': a strong outer surface of cyanoacrylate (a superglue-like substance) is carefully applied, and this process also enhances the colour. Ultimately, the print provides a relatively robust material record: a remediation, a literal 'hard' backup of the digital file and a comparatively durable, if imperfect, replica of the fragile object. It has been on display in an exhibition about old and new print processes, and handled at talks and conferences. It now spends most of its time in the conservation studio: a document sitting in close proximity to other conservation documents, and an object of some curiosity among visitors to the archives.

Decoding, disclosure and deceit

Ian Dawson's collaborative project *The Wanderer's Nightsong II* addresses head-on issues of remediation, authenticity and the complex status of computer-generated artefacts, as art objects are copied via 3D printing and shown next to the originals. As is explained in the press release for the exhibition, 'the work of five artists will begin to dialogue with bastard versions of themselves, to communicate across this space' (Dawson 2015). And a pertinent question is asked about the nature of the 3D print: 'Just as the photograph altered our perception of reality to include the blur – what will the 3d scan and print do? What new forms of blindness will they induce?' (ibid.). Our 3D print (Figure 10.4) materialized from an application of image to points in space: it is a blur, a crudely thickened invention, a hollow surface, a skin with no body inside. It therefore voices the questions about replication, representation and documentation that are asked by Steyerl at the beginning of her essay. In the end, Steyerl (2017: 205) upholds 3D as 'an uprising of images, against an architecture of representation that holds them in servitude and subjects them'. This is fighting talk, and it must be set against the rather less upbeat, though not contradictory, concluding lines of the *Wanderer's Nightsong* text: 'There is a point in the not too distant future when computational power overtakes that of the human brain, a point

Figure 10.4 3D print of the Wellington bundle, 2018. © Jane Birkin.

when the baton for our evolutionary progress is fully transferred' (Dawson 2015). As 3D printing has now become an everyday and very sophisticated mode of production, the 'imagination' that Flusser (2000: 8) deems necessary for both the making and the decoding of 'images' is fast becoming redundant. We do not need, for instance, to project these surfaces through space and time and back again; it is done for us: 3D surface to 2D surface, and back to 3D – with the fourth dimension always already present.

The complex surface of the Wellington bundle prompts feelings about the letters it contains, and ultimately it probes tensions that are present within the archive itself. For example, it questions whether surface is truly important within the information-oriented institution of the archive, outside of the considerations of conservators and artists and the romantic illusions of a public looking for relics. Ingold (2017: 102) argues that surfaces are important 'for what they potentially open up, and for what they disclose. But they are also important for what they hide, and for the deceit that they can practise on us.' We might easily here term the original bundle as connoting disclosure, and the printed surrogate as an object of deceit; but this would be a crude classification, a deceit in itself, one that denies the complexities of the two. After all, it is the original object that is closed and secret, because of its own material condition, and because of the condition of the archive. The surrogate, on the other hand, is out there in the world; it is robust, open and with nothing to hide.

List of Figures

Chapter 11
Experience, Poverty, Transparency: The Modern Surface of Interwar Glass

Freyja Hartzell

Glass is magic. Not quite solid, not quite liquid, it is a shape-shifter: transforming itself, and – when penetrated by light – transfiguring the world. A material ever in flux, it appears now fluid, now frozen, and has been exploited practically and politically for both its material and its immaterial qualities over the course of its history. This essay examines glass objects and architecture produced in twentieth-century Germany as they not only manifest but mobilize the turbulent, shifting values and meanings of interwar modernity. Beginning in 1914 with the coloured glass utopias of science fiction writer Paul Scheerbart and his collaborator, the visionary modernist architect Bruno Taut, it traces glass's subsequent, systematic dematerialization: its evacuation of colour, texture and expressiveness, and its reorientation during the Weimar era (1919–33) and the Third Reich (1933–45) towards purity and colourlessness – utter transparency – in the hands of avant-garde architects such as Ludwig Mies van der Rohe and Bauhaus-trained designers such as Wilhelm Wagenfeld. The glass surface is elusive: it is at times opaque, denying access and resisting penetration – 'standing against' us in Martin Heidegger's sense of the object (*Gegenstand*), and thus, in its reflectivity, throwing us back upon ourselves to perform our own process of reflection (Heidegger 1954). But during the first decades of the twentieth century, 'pure colorless glass', as Wagenfeld termed it in a 1963 speech, became increasingly 'invisible'. Technology transformed it into the sheerest of screens – a surface whose perverse desire was to be wholly disregarded. Only 'light and shadow reveal its form to us', Wagenfeld (1963: 2–3) concluded; 'we feel the glass, but we see through it as if it did not exist'.

Since the advent of the Roman Empire (27 BCE), when blown-glass vessels became ubiquitous, the functional merits of transparency in utilitarian glass have been apprehended and pursued – only impurities of materials and inconsistencies in process preventing its perfect achievement. From the time of its discovery around 3500 BCE through to the middle of the nineteenth century, glass occupied the various material categories of 'object' and 'surface' rather than the immaterial space of the almost-imperceptible screen. The degrees of purity and colourlessness necessary to truly transform glass from a surface to be looked *at* into a functional, although invisible, screen to be looked *through* were attained not too long (in the terms of glass's five and a half millennia) before Wagenfeld's 1963 speech; and it was this transformation – and its implications for utilitarian objects made from glass – with which he was deeply concerned, in both thinking and designing, for much of his career. The pivotal figure in this process was the German glass chemist Otto Schott (1851–1935), who, in addition to realizing many other innovations in glass technology, discovered in 1884 that one could alter the optical properties of glass by changing its chemical make-up. Although the twentieth-century desire for 'invisible' household glass was for the most part aesthetic, the technology that inspired and enabled it had been developed and adapted for use in the fabrication of lenses for microscopes and telescopes – surfaces designed for the express purpose of being disregarded, *seen through*. Wagenfeld's un-surface – his 'pure, colorless glass' – was rooted in the scientific precision of late nineteenth-century optical science.[1]

But Wagenfeld's preoccupation with the material nature of glass transcended the purely scientific: his fascination was not simply with what glass could do, but with what it could do *to* those who worked and interacted with it. 'Once you fall for this strange material', he contended in 1963, 'you remain for the rest of your life under her spell' (Wagenfeld 1963). Over the course of more than five millennia, Wagenfeld's 'strange material' had been regarded as magical, not merely due to its appearance, but much more in terms of its affect. Whether it was the magic of sparkling, candy-like, coloured gemstones, the divine mystery of the Holy Spirit made visible as light passing through a stained-glass window or the electric enchantment of Cinderella's glass slipper and Snow White's glass coffin, glass both embodied and evoked wonder. Glass was, in essence, successful alchemy: base materials of sand and ash were transformed not into gold, but into a substance far more enchanting: one that surpassed gold in its capacity to pass for a variety of natural products far more precious than itself. Wagenfeld (ibid.) argued that no other material was so 'fantastical, so improbable': 'One would like to believe that glass was not invented at all, but had, like rock crystal, always existed, and was simply discovered, once upon a time.'

But the alchemical magic of glass flows from its nature as a material in motion. Not a true solid like rock crystal, but a supercooled liquid, its essence is flux. Wagenfeld (ibid.: 2–3) understood both glassmaking and glass itself as phenomena of transition: 'We know that glass is artificially made and composed of various natural materials, which become fused with one another in fire. But this fusion is also a transformation. A transformation of what is naturally occurring to that artificiality which is glass.' As the common root of 'transition', 'transformation' and 'transparency' suggests, glass has been understood over the centuries as perpetually in a state of 'passing': not simply morphing, but imitating – 'standing in' for other materials and ideas. In fact, the ancient Greek *hyalos* refers variously to naturally occurring crystal and human-made glass; here 'glass' is defined not materially but functionally: it is that which permits light to pass through – that which strives to negate its own presence, efface its own self, in the service of transparency. The arguments that follow hinge on glass's dual nature – as both substantive surface and insubstantial screen. Are its affective capacities (its objecthood or thingness) bound to its perceptibility – its colour, texture, the degree of its opacity? Or is glass most at home, most itself, when least 'there'? Is glass for looking *at* or looking *through*, and how does the opposition of its material manifestations – and our perceptual modes – inform its cultural, social and political life in the first decades of the twentieth century? If the impulse towards transparency, as Wagenfeld believed, is essential to the nature of glass, does this nihilistic drive darken its fairy tale?

Erase the traces: clean slates and modern barbarians

In *Weimar Surfaces: Urban Visual Culture in 1920s Germany*, historian Janet Ward (2001: 62) terms glass 'the medium par excellence of the clean, clear surface'. This reflects and summarizes the strain of glass discourse that is most familiar to the history of modernism and was epitomized in remarks on the subject by German-Jewish expatriate and cultural critic Walter Benjamin in 1933. These appeared in an essay titled 'Experience and Poverty' in a Czech journal, less than a year after the burning of the Berlin Reichstag and the assumption of political power in Germany by the National Socialist German Workers' Party. 'It's not for nothing', Benjamin (2014: 217) writes, 'that glass is such a hard, smooth material to which nothing can be fixed. A cold one and a sober one, too. Things made of glass have no "aura". Glass is generally the enemy of secrets. It is also the enemy of possession.'

Benjamin was writing in 1933 about a particular generation whose 'experience' (*Erfahrung*) up to that point had been unlike the experience of any previous generation. The unique experience of the generation that came of age between the two world wars, Benjamin argues, was actually a *negation* of experience, a kind of emptying out of established middle-class life, and a liquidation of the environments and objects that enabled it – an experiential and material impoverishing. Benjamin (1999: 731–2) asserts:

> Experience has fallen in value amid a generation which from 1914 to 1918 had to experience some of the most monstrous events in the history of the world. … Wasn't it noticed at the time how many people returned from the front in silence? Not richer but poorer in communicable experience? … For never has experience been contradicted more thoroughly: strategic experience has been contravened by positional warfare; economic experience, by the inflation; physical experience, by hunger; more experiences, by the ruling powers. A generation that had gone to school in horse-drawn streetcars now stood in the open air, amid a landscape in which nothing was the same except the clouds and, at its center, in a force field of destructive torrents and explosions, the tiny, fragile, human body.

The multiple cataclysms rippling out from the mechanized mass destruction of the First World War had 'contradicted' all pre-war experience, emptying it of its prior contents – or meanings. However, these systematic eviscerations had not simply impoverished everyday experience; they had rendered it transparent as well. The 'tiny, fragile, human body', which before the war had been draped, padded and swathed in the protective tissues of private bourgeois existence, was now, according to Benjamin, utterly exposed to the outer world of public events – standing 'in the open air … in a force field of destructive torrents and explosions'. In Benjamin's (ibid.: 732) view, the entirety of *culture* – and especially visual and material culture – had been bankrupted. In the post-war wasteland there was nothing left to hold onto, nothing to hide. And the impoverished – naked – generation that inhabited this barren landscape was also the first to experience transparent glass architecture and mass-produced glass objects as features – and even markers – of modern domestic life.

Although utilitarian glass vessels had been commonplace since the Roman Empire, and although iron-and-glass exhibition halls, train stations and department stores were by 1900 already familiar landmarks of the modern urbanscape, it was not until the first decades of the twentieth century that technological developments in glass manufacture confronted the public with the simultaneous (and paradoxical) presence and absence of the pure, colourless glass surface on both miniature and monumental scales. In addition to Schott's late nineteenth-century innovations in the chemical purification of glass, in 1901, Belgian glassmaker Émile Fourcault revolutionized architectural glass construction by inventing a window-making machine that drew a flat sheet of glass five storeys directly up from a vat of molten glass. Prior to this, sheet glass had been made by opening and flattening large blown-glass cylinders. But the glass façade or 'curtain wall' – first attempted by architects Walter Gropius and Adolf Meyer in the design of the Fagus shoe factory in Alfeld, Lower Saxony (1914), and achieved to greater renown by Gropius with the 1925 Bauhaus building in Dessau – hinged not simply on shifting the building's load-bearing

Figure 11.1 Walter Gropius, Bauhaus, Dessau, *c.* 1925. Photograph: author (2003). © 2019 Artists Rights Society (ARS), New York/VG Bild-Kunst, Bonn.

capacity from its outer walls to its interior structure, but on the new availability, in quantity, of large expanses of flat, transparent glass (Figure 11.1). Gropius ([1937] 1989: 43–4) wrote that this 'New Architecture throws open its walls like curtains to admit a plenitude of fresh air, daylight, and sunshine'.

But for Benjamin, transparency was, especially in its relation to the Great War, an experience of rapid cultural disillusionment, a radical evacuation of culture. This cultural emptying amounted, in Benjamin's (1999: 732) view, to a 'new kind of barbarism':

> Barbarism? Yes, indeed. We say this in order to introduce a new, positive concept of barbarism. For what does poverty of experience do for the barbarian? It forces him to start from scratch; to make a new start; to make a little go a long way. ... Among the great creative spirits, there have always been the inexorable ones who begin by clearing a tabula rasa.

Starting 'from scratch' was, for Benjamin, the precondition of modernism. He counted among the proponents of this fierce, ruthless modernism figures now familiar to the history of art and architecture, including the painter Paul Klee, and the architects Adolf Loos and Le Corbusier. But the modernist most heroic to Benjamin was neither artist nor architect, but the Berlin poet, science fiction writer and utopianist Paul Scheerbart (1863–1915), who 'greeted the present with greater joy and hilarity' than any other modernist 'barbarian' (ibid.). Scheerbart greeted the present with glass and, through it, envisioned the future. Benjamin (ibid.) writes that Scheerbart housed his fictional characters 'in adjustable, movable glass-covered dwellings of the kind since built by Loos and Le Corbusier'. Writing almost twenty years after Scheerbart's 1914 *Glass Architecture* – a part-practical, part-visionary manifesto in 111 propositions for the uses of glass in modern domestic architecture – Benjamin enlists Scheerbart in his utopian vision of glass as the great cultural emptier and social leveller, the 'enemy' of secrets and possessions. 'Do people like Scheerbart dream of glass buildings', Benjamin (ibid.: 734) asks, 'because they are the spokesmen of a new poverty?'

According to Benjamin, glass's primary social function is its capacity to dissolve the opaque, overstuffed late nineteenth-century bourgeois interior, which he repeatedly (albeit affectionately) lambasts in the 1930s:

> If you enter a bourgeois room of the 1880s, for all the coziness it radiates, the strongest impression you receive may well be, 'You've got no business here.' And in fact you have no business in that room, for there is no spot in which the owner has not left his mark – the ornaments on the mantelpiece, the antimacassars on the armchairs, the transparencies in the windows, the screen in front of the fire. A neat phrase by Brecht helps us out here: 'Erase the traces!' is the refrain from the first poem from his *Lesebuch für Städtebewohner* [Reader for City Dwellers, 1927].
>
> (Ibid.)

Bertolt Brecht's images of 'traceless' modern living earn him a place among Benjamin's 'good barbarians'. Unmoored from social convention and unfettered by familial obligation, Brecht's young city dwellers live relentlessly transparent and decidedly uncosy lives, dwelling no longer within the comfortable apartments of their forebears, but now in all senses *without*. In the outer 'force field of destructive torrents' they start 'from scratch', making every place – and no place – their 'dwelling'. Their defiantly rootless, austerely anonymous, obsessively mobile life is as clear and as sharp as glass. The pre-war dweller has been blasted by the explosive, relentless realities of war into a brilliant and bristling outer life – the crystalline cladding of the modern barbarian.

Glass culture: moral exhibitionism, transparent anatomy

Benjamin (1999: 734) wrote in 1933 that between the Bauhaus 'with its steel' and Paul Scheerbart 'with his glass', modern interiors 'in which it is hard to leave traces' had finally been realized. For Scheerbart, writing in 1914, rather than simply representing the blankness of Benjamin's rhetorical tabula rasa, the material of glass itself acted as an actual clean slate:

> We live for the most part in closed rooms. … Our culture is to a certain extent the product of our architecture. If we want our culture to rise to a higher level, we are obliged … to change our architecture. And this only becomes possible if we take away the closed character of the rooms in which we live. We can only do that by introducing glass architecture, which lets in the light of the sun, the moon, the stars, not merely through a few windows, but through every possible wall, which will be made entirely of glass – of colored glass.
> (Scheerbart [1914] 2014: 26)

In its call for a new openness and permeability of 'closed rooms', Scheerbart's pre-war referendum on the conventional opaque interior, coupled with his prophetic vision for a modern culture transfigured through the domestic application of glass, both prefigures Gropius's 1937 image of permeable glass walls opening 'like curtains' and resonates with Benjamin's 1933 critique of the cosy bourgeois interior of the 1880s, swathed in its multitude of covers and casings. For all three, glass was not simply a modern domestic convenience, but a utopian substance: it was the clear and clarifying antidote to the close and cloying interiors that had lulled the private dweller of the nineteenth century into dangerous complacency in regard to public events. In 1929, Benjamin ([1929] 1978: 180) wrote that 'to live in a glass house is a revolutionary virtue par excellence … a moral exhibitionism, that we badly need'. Glass was a Marxist material: it not only erased Brecht's sedentary bourgeois 'traces', but dissolved the blinding barriers between private domesticity and civic responsibility. Glass offered political transparency and collectivist collaboration.

Modern glass architecture and objects positioned themselves as politically progressive. Architectural historian Anthony Vidler (1992: 217) has referred to modern glass architecture – opening buildings up 'like anatomical models, its walls hiding no secrets' – as the 'very epitome of social morality'. As if to entrench modern glass in its political stance, not long before Benjamin published 'Experience and Poverty', the National Socialist German Workers' Party had become the public enemy of a modern glass icon. In September 1932, the Nazi-dominated city council of Dessau forced the closure of Gropius's brazenly transparent steel-and-glass Bauhaus and the progressive workshops that occupied it.

The same transparency that Benjamin and the Bauhaus hailed as social and political liberation, the Nazis decried as un-German and degenerate. But as historian Paul Betts (2004: 63) has shown, 'while the Bauhaus itself was

Figure 11.2 Wilhelm Wagenfeld, Kubus storage containers, *c.* 1938. Vereinigte Lausitzer Glaswerke, AG. 1. Museum Associates/LACMA/Art Resource, NY. © 2019 Artists Rights Society (ARS), New York/VG Bild-Kunst, Bonn.

dramatically closed in 1933 as an unwanted scourge of "cultural bolshevism", several Bauhaus products ... stayed in production through the early 1940s, complete with the supposedly taboo Bauhaus moniker'. Indeed, the Nazis continued to support the work of several Bauhaus-trained artists, including that of glass designer Wilhelm Wagenfeld. In 1938, Wagenfeld designed for the Lausitz Glassworks a set of stackable food service and storage cubes whose capacity for reconfiguration into a seemingly limitless variety of architectural massings likened them to a set of children's building blocks (Figure 11.2). The Kubus containers' manufacture from cheap, durable, heat-resistant borosilicate glass (a product pioneered by Schott in the 1880s) set them in motion from refrigerator to table in households across the Third Reich. Although loved by the Nazis, Wagenfeld's cubic vessels – their cargo laid bare – mimicked, in miniature, the hated Bauhaus itself, with its curtain wall exposing controversial ideological 'contents'. Contrary to Benjamin's theoretical musings, then, transparent glass was, practically speaking, anything but a clean slate; it was, by the time of Benjamin's 1933 essay, a container filled with cultural and political meaning.

Strange fire: coloured glass and modern magic

Although both Scheerbart and Gropius understood glass as the key to a new culture of permeability between the progressive domestic interior and the fast-paced modern world outside, the glass surfaces they envisioned were radically opposed, manifesting each man's distinct apprehension of what glass was and what it could – or should – do. While Gropius's glass wall was a passive curtain, meekly opening to allow the penetration of light and air, Scheerbart's glass wall was an active surface, affecting the character of both the light that passed through it and the dweller upon whom that light fell. Between 1889 and his death in 1915, Scheerbart published about thirty pieces of writing on glass – most of them fiction. Architectural historian John A. Stuart (1999: 65) characterizes Scheerbart's project as 'fusing the spiritual ideals of materiality to the process of architectural construction through narrative'. Central to Scheerbart's narratives was the experience of transfiguration, most notably embodied in *Lesabéndio*, published in 1913. Lesabéndio, a visionary astronomer living on the fictional 'double star', Pallas, convinces his fellow Pallasianers to construct a 44-mile-high tower that will enable the connection of the star's two halves. In this process, the peaceful, anarchist community of Pallas undergoes a painful transformation, and Lesabéndio himself is transfigured into colour energy, finally dissolving entirely into light (Scheerbart [1913] 2012).

Scheerbart's contemporaneous friendship with the expressionist architect Bruno Taut prompted him to embark upon *Glass Architecture*, his first non-narrative, technical proposal for employing glass to revolutionize the experience of modern culture. Taut had just received the commission to design his Glass House – a faceted, crystalline pavilion constructed from a variety of glass products fabricated by the German Luxfer Prism Syndicate – for the 1914 German Werkbund exhibition in Cologne. That same year, Scheerbart published *Glass Architecture* along with what would become his best-known work: *The Gray Cloth and Ten Percent White: A Ladies Novel*. The two have since been understood as companion pieces: *The Gray Cloth* fictional, and at times irreverently ironic, and *Glass Architecture* near-ecstatic in tone and utopian aspiration, but also strikingly pragmatic. The futuristic *Gray Cloth* imagines the implementation of modern glass architecture through the story of the Swiss glass-architect Edward Krug, who travels by airship to erect colourful glass structures around the world. Krug marries a young Chicago organ player, Clara Weber, on the condition that she agrees to wear only 'gray cloth with ten percent white', so as not to upstage his vibrant glass buildings. While Clara agrees to wear this muted clothing, others attempt to dissuade her; eventually, however, she emphatically commits to her grey fate, supporting her husband's position through her attempt to 'seduce people into using colors' in their built environment (Stuart 2001). In Scheerbart's *Gray Cloth*, the roles of subject and object are reversed: Krug's glass buildings are aesthetically dynamic, loud, strident, while his wife's person is intentionally subdued, muted, still.

Twenty years later, in 'Experience and Poverty', Benjamin would take up *Lesabéndio*'s image of a society radically transformed through the experience of cataclysm, manifested in the dissolution of boundaries through the penetration of light – through transparency. And before Benjamin's tabula rasa, Scheerbart's architectural fantasies in *The Gray Cloth* had promised a new beginning. But Krug's coloured glass was hardly Benjamin's uncompromising 'enemy of secrets'. It was playful, mysterious and seductive, revealing its effects neither immediately nor transparently, but gradually – through coloured filters. Revere him as he might, Benjamin's appreciation of Scheerbart was at the same time a significant reinterpretation of the latter's utopian vision.[2] At odds with Scheerbart's flamboyant, idiosyncratic and frequently absurd glass constructions was Benjamin's conviction that glass was an auraless, 'impoverished' substance with which to build a stark, naked modernity (see Bletter 2014: 127).

While colour, texture and ornament characterized Scheerbart's utopian vision for modern life, his glass utopia was as practical as it was fanciful: in his 1914 *Glass Architecture*, the theoretical treatise proposing glass as the modern architectural medium par excellence, Scheerbart inveighed against traditional building materials such as brick and wood for their incompatibility with modern doctrines of hygiene. Scheerbart's highly original yet quintessentially modernist strategy was to collapse what design reformers of his day referred to as 'applied' ornament into the very structure that supported it. Like Gropius, Scheerbart's ambition was to build with glass – just as modern architects of his own day were already building with iron and ferroconcrete. But rather than stripping his aspirational glass façades of decoration, Scheerbart infused decorative delight into the very substance of his structures by dream-building with faceted, coloured glass.

Scheerbart dedicated his 1914 *Glass Architecture* to Bruno Taut, who, in the summer of the same year, experimented with Scheerbart's propositions for a modern 'glass culture' in the form of his Glass House at the Deutscher Werkbund exhibition in Cologne (Figure 11.3). Circling the structure were short aphorisms composed by Scheerbart. Some proclaimed the hygienic benefits of glass architecture – 'Vermin are not nice / But they'll never come into a glass house' – while others inculcated glass's transformative virtues: 'Colourful Glass Destroys Hatred' (my translations). Taut's pavilion materialized Scheerbart's glass dreams in a host of reflective and translucent materials: walls were alive with colour and movement, surfaces saturated with depth. Taut's structure was not to be looked through, but to be looked at – to be optically experienced. Taut narrates a tour of the interiors, where walls constructed from Luxfer's thick, prismatic glass bricks were illuminated by both natural and electric light. 'An opening in the floor looks

Figure 11.3 Bruno Taut, Glass House, Werkbund Exhibition, Cologne, *c.* 1914. From *Deutsche Form im Kriegsjahr: die Ausstellung Köln 1914* (München: F. Bruckmann, 1915).

into the basement', Taut (2014: 103) writes, 'glass staircases lead downward … to where a waterfall bubbles forth. … This cascade is made of ornamental and plate glass mounted on … unpolished glass and is lit from behind with electric lamps. The cascade's downward trail leads to a purple fabric-lined niche with a screen, upon which rhythmically shifting kaleidoscopic images are projected, reminding the viewer of childhood'. With an aesthetic antithetical to that of the smooth, transparent sheet glass that constituted the first curtain wall at Gropius's contemporaneous Fagus factory, Luxfer's textured, translucent glass bricks, their depths animated by internal flickerings, fulfilled Scheerbart's dream of building with glass. This glass was incapable of self–effacement; rather than opening like a curtain at the visitor's behest, it enveloped and suspended her attention with refracting motion and reflecting light.

The Werkbund exhibition was cut short by the outbreak of the First World War in August 1914, and Paul Scheerbart met his untimely death in October 1915. Throughout the war years, as building industries ground to a halt and construction materials became increasingly scarce, Taut occupied himself with modest – though still visionary – projects. In 1917, he began work on a folio of colourful watercolour drawings, published in 1919 as *Alpine Architecture*.[3] This collection of images, carefully annotated with lyrical descriptions and affirmations in the rounded, fluid characters of the artist's own hand, constituted the first of Taut's illustrated works relating glass architecture and alternative visions for society. Unconfined by the realities of construction, Taut was able to respond more directly to Scheerbart's propositions for a glass culture than ever before: Taut's 'crystal house' in the mountains was to be 'built entirely of glass – coloured'; the 'jewel-like glass architecture' of Taut's Cliff-Cathedral (pictured in *Alpine Architecture*'s plate 11) 'shines forth into the mountains and the heavens'. For Taut, as it had been for Scheerbart, the significance of glass architecture was not its cool, spiritual perfection, but the warm, social radiance this necessarily produced. Like Gothic cathedrals, Taut's pacifist constructions were collective projects – designed to transform the social body that built and used them (Sharp 1972: 122–6). In *The Dissolution of Cities*, published the following year, Taut depicts an entirely anarchist society, in which not only cities, but all institutions – governments, schools, marriage – have been dissolved. Here, glass structures are entirely flexible and reconfigurable. And in this flexibility, they are also entirely communal.

Taut's vision for the social relevance of glass architecture, its utopian potential, and specifically the childlike, imaginative sense of play that coloured glass evoked in those who worked with it, all combined in another speculative project he worked on during the same period: 'Dandanah – The Fairy Palace' (1919–20), a set of coloured glass building blocks for children, manufactured, like the translucent bricks for his 1914 Glass House, by the German Luxfer Prism company (Figure 11.4). Taut developed Dandanah while involved in the Crystal Chain, a group of expressionist architects and artists, including Walter Gropius, who maintained an ecstatic correspondence around the themes of utopia and glass architecture between November 1919 and December 1920. The children's blocks embodied Taut's and Scheerbart's conception of glass as magical, mobile and *alive*.[4] When light passed through the coloured blocks they seemed to catch strange fire: Crystal Chain member and architect Hermann Finsterlin described his daughter playing with the glass blocks by candlelight in a darkened room; Taut, himself, writing under the pseudonym 'Glass' in a letter to the Crystal Chain in April 1920, observed with a child's rapture the mysterious, sensuous, elusive and *animated* effects that resulted from 'playing' with a piece of coloured glass:

> I have here on my table a thick piece of yellow glass. Heavy as a building brick, constantly changing in appearance. Certainly, its prismatic form is constant, but there is an ever-changing life in it. It's simply fantastic what effects the light produces, and yet within a fixed form. The vessel of the new spirit that we are preparing will be like this.
>
> (Whyte 1984: 84)

Figure 11.4 Bruno Taut, Dandanah children's block set with instruction cards, *c.* 1920. Pressed glass, printed cardstock. German Luxfer-Prism Syndicate. Canadian Centre for Architecture. Gift of Phyllis Lambert.

The Dandanah blocks were spiritual vessels in the most speculative sense: while they facilitated an intense, apparently almost-hypnotic experience of play, they also provided scope for the development of a constructive imagination during the interwar period. Dandanah's sixty-two geometric blocks in red, blue, yellow, green and colourless glass delivered the dynamic, prismatic experience of the experimental Glass House into small hands. The child who played with Dandanah, modelling her fantasies on six instruction sheets that encouraged the construction of massive, monumental, colourful buildings alive with light, was what Taut called an 'imaginary architect' (Whyte 1984: 19). Dandanah was a pedagogical tool, not simply in its function as a toy for developing spatial awareness and motor skills in young children, but, more profoundly, as a means by which children could discover the utopian potential of glass architecture, and so become transformative forces for a better future world. The glass 'vessel of the new spirit' that Scheerbart, Taut and the Crystal Chain correspondents were 'preparing' between 1914 and 1920 was neither cold, sober nor transparent; it was instead alive with light and dancing with colour: it was material, tangible, *sensual* – 'like this'.

Precise crystallometry: ice, glass, hygiene

In *Alpine Architecture*'s plate 10, titled 'Snow, Glacier, Glass', Taut envisions 'snowfields in regions of eternal ice and snow … built over and decked with embellishments in the form of planes and blocks of colored glass', as if some gigantic child had taken her Dandanah blocks on holiday in the mountains. Taut's glass blocks, designed to facilitate connections between creative play and speculative architectural thinking, were at the same time individual, self-contained objects – sleek and lapidary, like brightly coloured gems or crystals. In their simultaneous fixity and flexibility, the Dandanah fairy blocks embodied the mysticism and mystique of the crystal, which, unlike human-made glass, was a growth of nature – one whose development and accretion could be observed most directly and immediately not in slow-growing, multicoloured minerals, but in the comparatively rapid crystallization of colourless ice.

The idea that frozen-water crystals – the material and structural constituents of snow and ice – could be animated by an inner force, externalized in active growth, may seem counter-intuitive. Thomas Mann explores the paradoxical animacy of freezing in *The Magic Mountain* (1924), the 700-page novel that he was writing while Taut was in the process of producing *Alpine Architecture* (1919) and *The Dissolution of Cities* (1920). In a chapter entitled simply 'Snow', Hans Castorp, the young engineer and patient at the Alpine sanatorium around which the narrative revolves, is both attracted by the frigid aesthetics of what Mann calls the 'fairy-tale' Alpine winter, and also instinctively intimidated and repelled by the encroachment of this cold-blooded, crystalline 'other'. Around the sanatorium, the 'few paths still passable were like tunnels, with snow piled man-high on both sides, forming walls like slabs of alabaster, grainy with beautiful sparkling crystals' (Mann [1924] 1995: 461). But the ice crystal's very life – its growth, activity and accumulation – spelled the increasing torpor of cellular organs and systems, and eventually the cessation of all life. In the snow chapter, Mann presents this inverse relation as deeply ironic, through Hans Castorp's inner reflections on the physical make-up of the snow:

> [It] consisted of myriads of water droplets, violently gathered up and frozen in manifold, symmetrical crystals – little pieces of an inorganic substance, the wellspring of protoplasm, of plants and human beings; and among all those myriads of magical stars in their secret, miniscule splendor never intended for the human eye, no two were alike. It was all the result of an endless delight in invention, in the subtlest variation and embellishment of one basic design: the equilateral, equiangular hexagon. And yet absolute symmetry and icy regularity characterized each item of cold inventory. Yes, that was what

> was so eerie – it was anti-organic, hostile to life itself; Hans Castorp thought he understood why the architects of ancient temples had intentionally and covertly built little deviations from symmetry into their rows of columns.
> (Ibid.: 471)

In the act of freezing, the water droplets not only turned from constituents of organic life to its 'anti-organic' opponents, but also (in Hans Castorp's estimation) from something fluid, amorphous and potentially flawed, to something symmetrical, regular, perfected – *designed*.

Over the snow chapter's course, Hans Castorp's fight to keep from freezing after losing his way in a blinding Alpine blizzard is concomitant with his fight against the snow's perfecting design. His flesh itself rebels against the ice crystals' inorganic precision: 'I do not intend, my stormily pounding heart does not intend, to lie down and be covered by stupid, precise crystallometry' (ibid.: 475). Like the 'poor rich man' in architect Adolf Loos's infamous 1898 polemic on the pathologically overdetermined *Gesamtkunstwerk* interior, Hans Castorp is oppressed – to the point of death – by the precision of design. But unlike Loos's rich man, Hans Castorp rails against death by design with the very beating of his heart.

Hans Castorp fears being discovered dead in the snow, 'with eyes turned to glass' (ibid.). Mann's opposition of Hans Castorp's 'stormily pounding heart' to the snow's 'precise crystallometry' – and to its human-made analogue, glass – points to a broader philosophical and historical antagonism between what Elana Gomel and Stephen Weninger (2004: 74) have called 'the dead beauty of the crystal and the messy vitality of the body'. Although Hans Castorp's animal body reflexively resists the snow's deadly designs upon it, his rational, modern engineer's mind is compelled by the snow's straightforwardness, aesthetic restraint, logical order and *hygiene*: 'The snow, a deep, loose, unembellished powder, played the same role here as yellowish-white sand did down below; both felt clean to the touch … neither was dusty, neither left a trace behind' (Mann [1924] 1995: 463). At the same time that the perfection or 'dead beauty' of ice and snow threaten the 'messy vitality' of the human body, the frozen crystals – 'cold and sober', like Benjamin's glass – offer a primal example of tracelessness, a natural blank canvas: 'a surface [sanatorium] guests found useful for drawing pictures or writing messages' (ibid.: 461). Snow provides not simply an expanse of pure, white ground, but a literal 'clean slate': annihilating all traces of organic life.

Frozen light: death by design

Wagenfeld saw in glass 'the magic of frozen light' (Scheiffele 1994: 12). The development of design in Germany over the course of the 1920s could be compared to a gradual freeze: glass was turning to colourless ice. Wagenfeld's

Kubus refrigerator containers – clear, sterile and destined for the cold – were the designer's practical tabula rasa for the modern German home. The icy cubes pictured a different vision of modernity in 1938 than Taut's fiery fairy blocks had done in 1920. Taut's blazing coloured glass had ornamented the frozen mountain peaks of his unruly Alpine landscapes. But Wagenfeld's colourless, transparent cubes seemed to have killed off not just microbes but ornament itself, presenting German housewives with squeaky-clean slates: smooth, sanitary – and see-through – surfaces. Hygiene was no longer just a function of design; it was an aesthetic.

Wagenfeld's cubes for the Lausitz Glassworks were not his first foray into the poetics of 'frozen light'. One of his earliest designs experimented with the transparent and reflective surfaces that would, by the 1930s, come to dominate the aesthetics of modern design. This was the now iconic glass and chromed-steel table lamp that Wagenfeld designed as a student at the Weimar Bauhaus in 1923–4.[5] 'The transparent glass of the base and shaft … transmits the light', writes design theorist Frederic Schwartz (2009: 138), 'the milky glass of the shade diffuses it; the chromed steel in the center reflects it back'. Although the lamp's cool, clean transparency interprets modern glass in a manner radically different from Taut's eccentric anarchist fire, Schwartz (ibid.: 138–40) argues that it, too, promotes an agenda of egalitarianism: 'The white light of Wagenfeld … is even, objective, diffuse, everywhere the same. It won't destroy hatred, but it will show it clearly.'

In 1925, the Bauhaus workshops relocated from the historic, picturesque town of Weimar to the industrial city of Dessau, where they were installed in Gropius's steel-and-glass temple to modern design. It was this new Bauhaus that was closed in 1932 due to withdrawal of support by Dessau's Nazi government. The rights to all patents, equipment and even the Bauhaus trademark were put in the hands of architect Ludwig Mies van der Rohe, who had in 1930 assumed directorship from the openly socialist Hannes Meyer, director of the Bauhaus since 1928 (the year of Gropius's departure). Mies reopened the Bauhaus as a private institution in Berlin in October 1932, but finally dissolved the school in August 1933 as a result of Nazi pressure (see Welch 1993; Schulze 1985: 186).

Shortly before his involvement with the Bauhaus, Mies gained recognition for his own glass house: the 1929 Barcelona Pavilion. This exhibition space, which represented Germany at an international fair for the first time since the First World War, was a modernist monument of translucent semi-precious stone, reflective chrome-plated columns, and luminous green, grey and white glass, as well as clear, colourless glass. In Barcelona, Mies reinterpreted the glass house, relying on the clean, sleek sheen of the surface, rather than its more mysterious, flickering depths. The German commissioner for the exhibition, Georg von Schnitzler, an executive at the chemical conglomerate IG Farben, wrote of Mies' new glass house in unmistakably utopian terms, but laced with a distinctly nationalistic flavour antithetical to Taut's and Scheerbart's more mystical, expressionist reveries: 'We wanted to be able to demonstrate what we want, what we can do, who we are, what we feel and see today. We want nothing but clarity, decency, honesty. … Here is the spirit of the New Germany: simplicity and clarity of means and intentions – everything is open, nothing is concealed' (Neumann 2013: 14). The commissioner's wife, Lilly von Schnitzler, in her review of the pavilion, drew upon the mystical tradition of glass and precious stones, linking this long-established glass rhetoric to a modern vision of German nationalism – or *Volk*. She envisions Mies' glass house emerging 'as if from a fairy tale, not of the *Arabian Nights*, but from an almost transcendentally inspired music of eternal space. … We Germans owe gratitude to Mies van der Rohe, for he has succeeded in casting our spiritual circumstances into form' (von Schnitzler 1929: 287).

Mies' transparent, translucent and reflective pavilion stood as a beacon of Germany's utopian modernism on the international stage: its shining surfaces reflected a gleaming glimpse of how not only Germans but all the citizens of the modern world might one day live in clarity, honesty and openness. And yet, as architectural historian Dietrich Neumann has pointed out, the meeting of beneficent modernist transparency and sinister nationalist ambition apparent in the Schnitzlers' responses to Mies' project was anything but coincidental. Although on the one hand drawn to avant-garde aesthetics, the Schnitzlers not only tolerated but actually cooperated with the Nazis during the Third Reich. More than this, Neumann argues, the 'minimalist clarity' of Mies' glass architecture was not only acceptable to conservative Germans like the Schnitzlers, but was identified (its opulence notwithstanding) as the most appropriate representation of their *Völkisch* egalitarianism – their populist promise for a more modern, more hygienic, more comfortable fascist Germany (see Neumann 2013: 15).

Equating the Nazis' hatred of the Bauhaus with a hatred of modernist transparency is not simply a mistake, but a dangerous misinterpretation of fascism's significantly complex relation to the aesthetic culture of modernism – or the modern surface. In fact, as architectural historians Barbara Miller Lane and Wolfgang Pehnt have shown, Hitler himself seems to have been torn between the need to align the Third Reich with an aesthetics of enduring greatness (what he termed the 'glorious past'), such as that identified in Greek art and architecture, and the desire to proclaim the 'new spirit' of National Socialism with architecture and design that would be perceived not just by the intelligentsia but, more importantly, by the German public as progressive or 'new' (see Miller Lane 1968; Pehnt 1983: 60). Even architect Albert Speer's notorious Lichtdom or Light-Cathedral – the stunning light sculpture designed for Hitler's rallies in Nuremberg from 1934 to 1938, consisting of 152 anti-aircraft searchlights, spaced at 12-metre intervals and directed at the sky to create a system of blinding vertical bars around the audience – exploited both antique, classical order and cutting-edge military technology.[6]

Miller Lane (1968: 189) has pointed out that, in his rhetorical efforts to blend tradition with modernism, Hitler adopted the slogan '*Deutsch sein heisst klar sein*' – 'to be German is to be clear'. Although deeply hypocritical, Hitler's phrase underscores the persistent connection of modernity to transparency – as well as the political force of their conjunction – even after the liberal, progressive values and projects of the Weimar Republic had been systematically suppressed by the new fascist regime. Just as the concept of 'clarity', the facilitator of both literal and figurative light, had informed the Schnitzlers' nationalist interpretation of Mies' Barcelona Pavilion in 1929, so clarity and light together played a significant role in Nazi propaganda, exemplified not just in Speer's spectacular pageantry, but also in the far humbler everyday objects of domestic design produced for the mass market in 1930s Germany.

Perhaps even more ironic than the Nazis' appropriation of the socialist rhetoric of transparency and light, however, was the fact that their subsidization of work by progressive designers originally developed during the 1920s finally made modern design generally affordable in the 1930s, enabling it to achieve that status of true 'mass design' for the first time in the Third Reich. And since metal, concrete and wood were increasingly requisitioned for weapons production, paper, ceramic and glass became important sources of revenue for the Nazi economy. These commodities 'quite unchanging in actual design – became a favourite repository of Nazi myths and fantasies' (Betts 2004: 49). In essence, the Nazis took modern design hostage, forcing it to play host to their parasitic populism.

But how much force was really needed to evacuate the political agenda from a modern design already vacant? Beautiful, vacuous modernist surfaces – conveniently devoid of 'contents' – beckoned the Nazis not simply with their slick, hygienic façades and dispassionate, martial order, but with their very lack of specific regional or historical style: a stylelessness that the Nazis apprehended, popularized and propagandized as 'eternal' form, material evidence of 'timeless German greatness' (ibid.: 67). Over the twenty years between one world war and the next, mobile, multicoloured, anti-utilitarian, anarchist glass had been blanched and frozen. Inside Wagenfeld's Kubus containers was no anarchist fire, nothing that stood for – or against – anything; his ice was neither mysterious, animated nor encroaching: it was simply *empty by design*. Motionless and mute, such defenceless 'eternal forms' passed into enemy hands.

Years later, in the 1980s, Wagenfeld reminisced about the Kubus storage vessels, designed in 1938 and produced during his years at the Lausitz Glassworks in the late 1930s and early 1940s:

> Some people came to visit us [at the Glassworks] once and said that, just like the Volkswagen factory [newly constructed and opened in 1938], a new refrigerator factory would be built, and for this they needed our set of refrigerator storage containers, sized it up, and declared it wonderful. We

> sold the sets in huge quantities. Some were shipped to Kiel, and some to Le Havre. I was tricked into believing they were for export, but it turned out they were for the German navy.
>
> (Wagenfeld cited in Scheiffele 1994: 221)

Whatever their true motives, the visitors to Wagenfeld's glassworks apprehended the sleek, mass-produced Kubus containers, with their practical, heat-resistant borosilicate glass and clever conformity to the technological advances of the modern kitchen, as fundamentally populist objects, capable certainly of serving a direct, functional purpose within the complex machinations of the approaching war, but perhaps even more importantly, of bringing simple 'joy' to the German people – akin to the Volkswagen or 'people's car', which Hitler had just celebrated at a 1938 rally as built 'for the broad masses ... to answer their transportation needs, and ... intended to give them joy' (Hitler cited in Parissien 2014: 119). But along with the cutting-edge cachet of temperature-safe containers whose contents one could see, Wagenfeld's avant-garde product enforced a new, military order in the German kitchen: to maintain the sparkling, clinical image of domestic hygiene that transparency exacted, every mark, every smear and every fingerprint had to be painstakingly removed from the pristine glass surface. Mass-produced glass did not 'erase the traces' – it invited them. Erasing them was done by hand.[7]

Transparency, for the Nazis, had surface appeal. It offered the perfect disguise for a regime sustained by propaganda and dependent for its effectiveness on darkness and deception. Designers such as Wagenfeld reversed, in a sense, Hans Christian Andersen's ironic tale of 'The Emperor's New Clothes' (1837), in which the emperor is tricked into parading naked through his own streets by swindlers passing for weavers, who purport to weave for him cloth so exquisitely fine that it is invisible. Modern designers had generated a real product – but, nonetheless, it was one that stripped *them* naked. Their new glass was a transparent product that, instead of deceiving their clients, swindled its designers themselves, exposing them to exploitation by sinister forces.

Industrial modernism's icy glass cubes were indeed objects on which it was 'hard to leave traces'; their designs had evolved through the desire for both material and conceptual cleanliness. But taking modern design's complex and ambivalent history within the context of interwar politics into account, the notion of a 'Bauhaus modernism' as simply 'empty' becomes problematic, because despite its consciously evacuated design, it could not be emptied of meaning. Glass was never a 'clean slate' – neither for modernity's 'good' nor for its 'bad' barbarians – but rather a palimpsest. Its stylistic vacuity meant something unique, irreplaceable – in the same way that a key witness is one upon whose testimony a case depends. Betts (2004: 72) writes that the progressive, modernist design object, understood as an active subject, or agent, in Nazi cultural politics, became a 'living witness of cultural rebirth, social reconstruction, racial victory, and private pleasure'. It meant something, in the way that an important hostage, held because of a specific, provocative cause or set of values, means something. And the self-effacing modernist object – designed to be transparent, empty and free – was an ironically easy hostage. It was, in fact, not just a hostage but a host – a vessel which, although originally invested with socialist transparency, became quickly infested with National Socialist hypocrisy.

The story Wagenfeld tells in 1980 of the 1938 Kubus subterfuge seems at best confusing, and at worst a possible retrospective fabrication. Ultimately, whether Wagenfeld truly believed he had been tricked into collaborating with the Nazis or not may be of little real consequence, as the historical record would seem to exonerate him through its silence on the matter (Manske 2012: 172). But perhaps the underlying question is not what Wagenfeld believed about the Kubus commission in 1938 – or in 1980, for that matter – but why he felt deceived. For had he not, like modern industrial design itself, already retreated, evacuated the public field? Why did the Nazis fear the glass Bauhaus, while desiring the glass blocks by a Bauhaus designer? Was it because the first was so difficult to empty, while the last was all too easy to fill?

Notes

1 For more on Schott's discoveries, see Corning Museum of Glass (1999) and Scheiffele (1994: 47–73).

2 Several scholars have pointed to Benjamin's particular reading of Scheerbart: see Stuart (1999: 69); Mertins (1996: 12); and Bletter (2014: 127).

3 Reprinted in Schirren (2004).

4 See Whyte (1984); the website of the Deutsches Museum, Berlin, http://www.deutsches-museum.de/en/exhibitions/materials-energy/technical-toys/tour/; and Juliet Kinchin's discussion of Dandanah (2012).

5 Dr Julia Bulk of the Wilhelm Wagenfeld Foundation in Bremen has brought to my attention that a 1999 court ruling by the Hanseatisches Oberlandesgericht in Hamburg established Wagenfeld as the single author/creator of the lamp. Karl Jucker contributed preliminary sketches for the design.

6 The Lichtdom is documented in the propaganda film *Festliches Nürnberg* (1937).

7 Gill Matthewson (2009) has discussed the relation between Benjamin's theorization of 'traces' and the marking of actual glass surfaces.

List of Figures

Chapter 12

On Genealogy of the Translucent Screen and the Rehabilitation of the Ephemeral: Post-cinema, Installation, Performance

Oksana Chefranova

The épistémè of translucency

Contemporary theoretical reflection and artistic practice have turned surface phenomena into conceptual figures – fold, skin, veil and membrane. In what follows, I explore one facet of the pervasive fascination with surface imaginaries: the translucent screen as it appears and multiplies across the visual art practices of post-cinema, installation art and performance. The proliferation of translucent screens is owed to the latest changing technologies of textile, glass and plastic, which affect various surfaces – physical and conceptual – of artistic production while inviting new image-making possibilities (see Kaltenbach 2004; Brownell and Casbon 2011). In parallel with a return of the ephemeral in design and technologies of surface, recent gallery and museum installations offer an abundance of diaphanous screens that, by experimenting with the elusive physicality of lightweight, semi-opaque materials and ethereal fabrics, foreground diffusion and gauzy blocking as creative strategies. Operating through layers, relays, back-and-forth dynamics, and the relations between light and matter and between viewer and screen, translucency simultaneously reimagines modernist transparency and engages with the recent preoccupation with the skin-like surfaces of bio-art, technologies of smart glass, and see-through computer screens.[1] I argue that this profusion of translucent screens acts as an expression of a post-cinematic condition, accompanied by a return of transparency-as-translucency in architectural discourse, and reveals translucency as one of the post-digital compensatory aspirations of the screen, evocative of the desire to overcome the superficiality of the digital image by looking through and beyond its surface. I also argue for utilizing translucency as a theoretical and operative category to address the collapse of borders and binaries, both material and categorical, in relation to the screen as an object and a concept. Dispelling the excessive binarism that surrounds surfaces of different media, the phenomenon of translucency allows for the mutability of surface imaginary, introduces a necessary ambiguity, and – flickering on the edges of categorical divides such as optical and environmental, depth and surface, reality and virtuality – signals the screen's protean capacity for connection, layering and liminality.[2] By contextualizing the study of translucency within a history of the screen from Daguerre's Diorama to contemporary visual arts, I suggest that translucency challenges and expands our notion of the screen, reshapes vision and decentres the subject. The translucent screen, both an aesthetic object and a theoretical concept, rechannels our thinking about the visual arts, cinema and new media in terms evocative of skin and atmosphere rather than technology and machine.

An expansive context of material practices and theoretical ideas has always surrounded the discourse of surface and the conditions of transparency/translucency/opaqueness across architecture, visual arts and performance. Modernity, as Anthony Vidler (1992: 217) proclaims, has been haunted by the myth of transparency with its 'ubiquitous flow of air, light, and physical movement'.

MODERN	POSTMODERN	CONTEMPORARY / METAMODERN
transparency	**opaqueness**	**translucency**
glass	surface	plastic
light	façade	light
clarity	palimpsest	skin
rationality	blocking	membrane
air	fold	seeing-through
airplane	simulacrum	ephemerality
radio	eclipse	diffusion layering
openness	superficial abyss	atmosphere
progress		blur vagueness
utopia		veil superimposition
		cloud
		meta-/trans-/post

Table 12.1 Modern, Postmodern, Metamodern

In parallel with the technologies of radio and the aeroplane, the modernist, panoptic, utopian glass of Paul Scheerbart and Bruno Taut promised to integrate nature and humanity within a unified transparent environment. Postmodernism eradicated modernism's aesthetics of transparency with the computer interface, proposing an apologia for surface and a reversal of clarity in favour of opacity, façade, palimpsest, simulacrum and 'superficial abyss' (Baudrillard 1988; see also Darley 2002: 58–77). In the contemporary era, defined by the pervasive 'post' prefix, transparency returns as translucency, growing into one of the most allusive tropes in contemporary architectural discourse, into a concept that holds contradictions, and, continuously opening up to new meanings, denotes, as Eve Blau (2008: 30) suggests, 'both visual and spatial connections, clarity and ambiguity, reflection and atmosphere, seeing and not seeing'. Air densifies into atmosphere, and translucency finds its metaphors in cloud, veil and membrane, and its relatives in 'deep surface' and 'light construction'.[3] In his psychoanalytic reading of architecture and architectural reading of psychoanalysis, Vidler postulates translucency as an uncanny repetition of modernist transparency, figuring both confirmation of transparency and its critique. Pointing to architect Rem Koolhaas's 1989 entry for the French National Library competition as a turning point, Vidler highlights 'shadowy presence', layered surfaces and three-dimensional volumes 'displayed ambiguously and flattened, superimposed on one another, in a play of amorphous densities' as the means by which transparency is converted to translucency, and ultimately to obscurity. In the uncanny moment of translucency, depth becomes surface becomes depth, allowing us 'neither to stop at the surface nor to penetrate it, arresting us in a state of anxiety' (Vidler 1992: 223). As a category defined by layers, networks and overlapping, translucency describes our post-perspectival perception aligned with the computer window, while on a more general level it exceeds a surface phenomenon and has consequences for subjectivity. As an example of paranoia and panic in architecture, translucency for Vidler reveals a new subject, one that is suspended in the condition of uncertainty, in 'a difficult moment between knowledge and blockage' (ibid.: 221). I suggest that in its return, translucency figures a new sensibility, a Foucauldian *épistémè* which in visual arts and architecture manifests itself through the performative potential of new materials and surfaces that, becoming more elastic and reactive, operate through blurring distinctions, perpetual flux and vagueness while shifting away from the architectural façade to architectural skin and membranosity, in correspondence with a re-emergence of the notions of dematerialization and ephemerality (see Imperiale 2000; Brownell and Casbon 2011; Murray 2010). Dutch cultural theorists Timotheus Vermeulen and Robin van den Akker (2010) – insisting on the difference between the oscillation that marks contemporary art and postmodern in-betweenness – highlight an emergent sensibility of layering that exposes tensions no longer describable in terms of the modern or the postmodern, but which 'must be conceived of as metamodernism'. The translucent mode, partaking in metamodernity, speaks to contemporary conditions of mediation with its ultimate impossibility of clarity while resonating with media theories' evocation of translucency 'as a medium's capacity for the incorporation and diffusion of other media' (Vasseleu 2015: 166).

Translucency and the screen

The present-day proliferation of screens and various modes of screening is mirrored in the increasing body of scholarly reflection that takes the screen as a central object of interest: from an archaeology of the screen that plays with screen continuity as a surface for images (Huhtamo 2004), to unravelling intricate etymological webs of its synonyms and connotations (Strauven 2012); from an ontology of the screen aiming to define what the screen is (Mitchell 2015), to mapping its different material conditions (Bruno 2014). The array of burgeoning studies stems equally from the desire to fill out lacunas in numerous screen histories as it does from the constant renewal and emergence of new screen formats and shapes that continually uncover fresh conceptual territories of enquiry. Dominique Chateau and José Moure (2016: 15) state that the screen remains 'a concept in progress' that has been opening an infinitely expanding field of study:

> From the screen as an object used to protect, obscure, or conceal, to the screen as an architectural and sculptural apparatus used to separate or divide space in a process of exclusion or delimitation, to the screen as a surface or a receptacle on which images are projected or displayed, to the screen as a metaphorical term or a site of mediation involving a relationship between what is shown and what remains under cover...

Giorgio Avezz (2016: 29) proposes a way of thinking about the screen not as an object with its own history – a premise of media archaeology – but 'as a function, or rather as a combination of functions'. The operation of screening in its many forms and practices reveals how the screen's meaning exceeds a mere repository of images. Similarly, Francesco Casetti states that 'screens are not only optical devices', a relatively late understanding that came into use in the first half of the nineteenth century. Rather, the screen has an environmental character, working 'as a prop to be used within and towards a space ... a filter, a divide, a shelter, a camouflage' (Casetti 2017). According to Casetti, the digital era fosters a return of these meanings, exposing the screen's double, back-and-forth dynamics – its movement from environmental meaning to the optical, before returning to the environmental. In her study of the materiality of surface, Giuliana Bruno also drifts away from the optical framework towards the haptic, reflecting the general shift in critical focus from optical and visual dimensions to other concerns. Bruno calls for materiality to be treated as 'the substance of material relations' rather than as a question of material itself, investigating the space of those relations and how they manifest themselves on the surface of different media. Proclaiming 'surface tensions' as a central condition of contemporary visual arts and architecture, Bruno builds one of her prevalent assumptions: any surface can become a screen when activated by projection. However, for Bruno (2014: 67), the effect of projection is never restricted to the purely visual but insists on the environmental – since the viewing experience takes place spatially, it is never purely 'looking-at' but rather 'looking-into', and the screen surface becomes metaphorically and literally inhabitable as a chamber and 3D space. While figuring the 'looking-through', an intervening medium, as a fundamental contemporary condition, the translucent screen transcends the optical-visual framework and gravitates towards a new embodied ecology and atmospheric affectivity.

Rather than understanding translucency as a level of transparency, I posit translucency as a separate category in relation to surface and to screen. Optically and materially, translucency means interaction between light and matter within a medium, and contrary to the transmission of light in transparency, translucency strives towards the diffusion of light, and thereby obscures. Transparency, by contrast, permits visual clarity. In transparency, the medium disappears; in translucency, there is always an entanglement of image, represented object and medium. Translucency is ontological to film: celluloid is translucent and holds an image imprinted by light, while projection is created by the diaphanous celluloid strip eclipsing light. The digital image suggests translucency by generating luminosity, a sense of light shining from within the image through its surface. Bruno (ibid.: 81) argues that the contemporary architectural interest in luminosity reconfigures the notion of surface in such a way that surface luminosity becomes an ambiguous space of translucency, neither clearly transparent nor fully opaque, but implying the façade as a veiling membrane. This notion of screen as veiling membrane reverberates throughout my examples. Diffusing light, the translucent screen embraces one of the screen's primordial meanings as a cover and protection from violent radiance, which makes it an instrument for thinking beyond projection, towards relations of layering and overlapping. Retarding the viewing process, forcing the viewer to notice the medium as well as its representations, the translucent screen exposes its lucid nature and imaginary openness while blurring the contradiction between visible and invisible. Since the Romantics, the diaphanous veil has alluded to the revelation of the invisible, while the capacity to diffuse light imbues the translucent fabric with magical and supernatural associations and the ability to generate visions and apparitions, imprinted in its etymology – the Greek verb *diaphaneite* may be rendered as 'you will allow [something] to shine through you' and 'you will cause [something] to appear' (Conlon 1980: 196). To see the invisible as manifesting inside the veil becomes crucial for spirit photography, in which materialization occurs inside the diaphanous veil, with ghostly faces or figures delineating themselves through and on a gauze-like cloth.

At the dawn of postmodernism, Colin Rowe and Robert Slutzky posit a categorical distinction within architectural theory between two transparencies. They differentiate between 'literal transparency', a material's inherent see-through quality associated with the modernist glass architecture whose simplicity leaves little to the viewer's imagination, and 'phenomenal transparency', which deals with spatial or volumetric organizations of overlapping, intricate layering and interpenetrating figures and planes, enabling visual complexity and multiple interpretations. Phenomenal transparency emerges for the theorists as 'a quality of spatial organization and relation', 'the simultaneous perception of different spatial locations' (Rowe and Slutzky 1963). Rowe and Slutzky borrowed the definition from the Bauhaus architect György Kepes (1944: 77, my emphasis), for whom transparency 'implies more than an optical characteristic: it implies a broader *spatial order*' of simultaneity and layering, based on the disturbance of figure–ground relations, on making it impossible to determine what might be considered to be the figure and what the field. The notion of phenomenal transparency permits the shift from optical aspects to environmental relations, working for Rowe and Slutzky as an operative concept for revealing distinctions between two orders of space: space *a priori*, offering itself as a container to be filled, and space that comes into being in the dynamic process of continually being constructed and articulated. In contemporary art discourse, Nicolas Bourriaud ([1998] 2002) offers a corresponding relational perspective with his notion of 'relational aesthetics', which takes as its subject the entirety of the dynamic environment, material and social, that embraces a discrete art object, and which insists upon the environment as a medium that produces relations. Similarly, I approach translucency as an activity of space and a relational category that embraces a multiplicity of material, perceptual and metaphorical interconnections partaking in the screen.

Tracing optical, spatial and atmospheric meanings of translucency, this essay aligns itself with the shift of focus from the screen as an object to multifarious relations between the screen surface and the environment that, materially and metaphorically, surrounds it. Instead of a chronological journey, it follows rhizomatic paths among evocations, allusions and correspondences towards a non-linear genealogy of the translucent screen, unravelling a web of formal and categorical relations that resonate across media, ideas and times. The genealogy created is reminiscent of Aby Warburg's *Mnemosyne Atlas*, which overlays different images on the singular plane of its black screen, causing the family relations between the images to come forward. The translucent screen brings together the material turn in the humanities, and the screen turn in film and media studies, with the luminous turn in art criticism and practice, as well as certain aspirations of the atmospheric turn. In what follows, instead of abstractly defining the notion of the screen, I fold several different screens that share a quality of translucency and are created by contemporary visual artists, who engage in their practices with themes and ideas similar to screen scholarship – surface, projection, vision, materiality, screening operations and screen environments. The first section describes how Stan Douglas's multimedia project *Helen Lawrence* (2014–16) negotiates the screen's surface in the digital era, and how its post-cinematic translucent screen stages the ontology of the digital image. In generating resonance between pre-cinematic visual culture and contemporary installation art, and bringing a media-archaeological dimension, the second section looks through the screen of the nineteenth-century Diorama to understand how, producing moving images without projection, it shifted the screen from the idea of surface to the idea of atmospheric environment. Finally, the third section addresses how in the installation-performance *Metakimospheres* (2015) the translucent gauze creates a second 'skin', which transpires as a screen for the projection of somatic data generated by invisible breathing, while exposing the corporeal and atmospheric affectivity of the contemporary screen.

Looking-through, the superimposition effect and the post-cinematic screen

'The eye stops at the screen', American avant-garde filmmaker Sidney Peterson declared, while Stanley Cavell (1971: 24) famously described the screen as an ontological barrier: 'What does the silver screen screen? It screens me from the world it holds – that is, makes me invisible. And it screens that world from me – that is, screens its existence from me.' The translucent screen proposes a counter-perspective to the cinema screen, which is materially opaque and omni-reflective but metaphorically transparent and disappearing at the moment of projection. Unlike the cinema screen, the translucent screen is pliable but neither submissive nor susceptible to disappearance and erasure. It imposes its presence on the viewer and emerges as a figure of screen doubleness, showing that the screen is always at least two things at once: offering reflection and filtering, the translucent screen is both the *screen on* and the *screen through*. Visual artist Stan Douglas, innovatively merging theatre, visual art, live-action filming and computer-generated imagery (CGI), works with such a polysemic and equivocal see-through screen in *Helen Lawrence*, whose optically penetrable surface does not capture vision, but allows a passage through and beyond projection, suspending the viewer between 'here' and 'there' (Figure 12.1). A product of post-cinema – this array of new image-making practices, technologies and conditions of viewing, in which there

Figure 12.1 Ava Jane Markus and Crystal Balint in *Helen Lawrence* during Brooklyn Academy of Music Next Wave Festival, *c.* 2015. Photograph: Richard Termine. Courtesy of BAM Hamm Archives.

is an excess of mediality – *Helen Lawrence* attempts a revision of the cinematic screen through a hybridization of post-dramatic theatre and live cinema, diffusing specificities of both the cinematic and the theatrical (see Denson and Leyda 2016). Staging the piece in the traditional space of theatre, Douglas replaces the theatrical curtain with a diaphanous scrim, behind which actors perform a film-noir story in front of a blue screen. The images captured by four cameras – three on moving dollies and one fixed at the centre, operated by performers onstage – are relayed to a computer, which aligns them with graphics and projects the CGI image to the scrim to create an augmented movie-like experience. The archivally reconstructed and digitally recreated historical locations are superimposed onto the real-life action, allowing the onstage actors to be transported into the simulated environment of 1948 Vancouver.

The screen subverts itself by exposing its reverse, repressed throughout film history – a blind spot in cinema but present from the beginning of screen media. This blind spot already fascinated viewers of early cinema, when the screen embraced its paradoxical nature as a surface that seemed at once penetrable and impenetrable, substantial and immaterial, transparent and opaque, as in an early viewer's memoir cited by Yuri Tsivian (2013: 155): 'The picture show made such an amazing impression on me that after the train scene I got up to take a look behind the screen.'

Similarly to Douglas's earlier installations, which referenced histories of cinema by reworking Alfred Hitchcock's *Marnie* or Dario Argento's *Suspiria*, *Helen Lawrence*'s inspiration comes from film noir of the 1940s and 1950s. The translucent screen literally acts as a veil, appearing to be a perfect noir screen in correspondence with *Helen Lawrence*'s commentary on this cinematic genre, which articulates itself via the veil epistemic based on the desire to see behind a surface, uncovering mystery and unveiling truth, and the dynamics of revelation and concealment. Complex in their manipulation with materiality of the filmic medium, Douglas's works aim at disturbing the film-viewing process with split screens, two-sided projections, loops, and manipulation with sound and editing (see Coulthard 2008). In *Helen Lawrence*, the artist explores one of his favourite operations – layering – engaging in a disorienting mélange of live space and filmed actors, disjunctions and uncanny doublings between figure and background, stage and screen, the profilmic and the image. Following his interest in the material aspects of the medium at hand, Douglas utilizes two old aesthetic forms, theatre and film, which become layered by the new medium of the computer. When the computer translates the real-time projections of the three-dimensional actors into two-dimensional screen apparitions superimposed onto their originals, the relationship between the actors and their translucent gigantic cinematic ghosts – figural analogy, contrast in size, and colour difference – reveals the multilayered coexistence of media. Projected on the spectral screen, a noir piece about hunting for traces and memories of the past, *Helen Lawrence* puts the media of theatre and cinema in a relation of doubling, so the experience of the *unheimlich* exceeds the narrative level and engenders the 'media uncanny' – a situation in which theatre and film reciprocally perform a 'return of the repressed' for each other.

The see-through creates in *Helen Lawrence* a visual situation evocative of the technique of cinematic superimposition and its effect of visual ambiguity and figures interpenetrating without optically destroying each other. When in *Cœur fidèle* (Pathé, 1923), one of the films of French avant-garde that favours superimposition as a linkage technique, Jean Epstein creates 'the fluid world of the screen' by superimposing two surfaces – the expressionless pale face of a woman and the sea – he enables different water movements, from light water ripples to dashing waves, to supplement the immobile mask with perpetual motion and unrest (see Epstein 2012). The superimposition effect here renders the human face a dynamic façade of forces, streams and tensions, and it is through this layering that the face comes into being as an expressive arena of interiority, yet not manifesting as an inner glow of the soul gleaming from the inside but composed by Epstein via the overlay and looking through the intervening medium of water. Peering past the volatile veil of water to the face, the viewer becomes attuned to the woman's emotions, while translucency enacts the epistemic uncertainty of looking-through as a necessary condition of vision – as defined by Maurice Merleau-Ponty, for whom vision is possible in the situation of optical eclipsing, veiling and looking-through rather than within unmediated transparency.[4] Interiority emerges as a new quality by means of cinematic superimposition as an aesthetic technique for bringing two disparate realities together. Jean-Luc Godard achieves a more radical effect of translucent visuality in his innumerable superimpositions of *Histoire(s) du cinéma* (1988–98), where layering of different images and media erupts the film's world by creating what – borrowing from poet Pierre Reverdy – Godard (cited in Morgan 2011: 138) defines as an image that 'cannot be born from a comparison but from a juxtaposition of two or more distinct realities'. Epstein's and Godard's superimposition operates at a limit of the cinematic form itself by suspending the sequential logic of cinema; it is this limit of cinematic sequentiality that Douglas tries to overcome with *Helen Lawrence*'s post-cinematic layering of cinema, with the order of simultaneity as a different articulation of time and space.

Mediating between the living actors and the projected figures with their uncertain corporeality and otherworldliness, the translucent screen thus produces the equivocal effect of two realities entering perception to foster a simultaneous presence. The mystified viewer is split between the stage and the screen, surrendering to Douglas's play with the idea of multiple representations, in which 'people are going to want to be able to have these experiences simultaneously, both the live and the cinematic' (Douglas cited in Byrne n.d.). Douglas's desire to dissect the experience of 'having two things that are contradictory or problematic somehow occupy the same space' results in the effect described by the artist as 'almost like cubism – we see things from different angles, and our perspective is always being disrupted' (Douglas cited in Battaglia 2015). Translucency as a quality of organizations and relations among different realities supersedes translucency as a mere quality of matter and surface.

Douglas's see-through membrane reveals, by hiding behind, the apparatus of image production: positioned in-between the camera and the computer/projector, the translucent surface exposes the screen's own partial nature, its inevitable inclusion into a larger constellation unfolding in the environment. But what it also divulges is the ontology of the digital image, its incomplete and composite nature, its constructedness as an object. The overlap between digital ghosts and real figures, the layering of a trace with its prototype, transpires into a commentary on the problematic indexicality of the digital image by showing that even if the referent is present in front of the camera, it is no longer sufficient for the manipulated assemblage of the digital image (see Doane 2007). Staging relations among different channels participating in the image, the translucent screen makes visible the pipeline of digital image production while exposing the ecology of the cinematic apparatus, i.e. a spatio-temporal assemblage of cinema's heterogeneous elements – camera-eye, projector, auditorium space, screen and spectator-eye, which according to apparatus theorists formed a coherent and ideologically powerful ensemble.[5] This theory famously acknowledged that technology disguises how reality is put together on-screen while seamlessly suturing the spectator into the cinematic process. As the translucent screen's material properties are more pronounced – not unlike the television screen, which also challenges this theory with the movement from projection to broadcasting – it rewrites the apparatus debate by anatomizing the solidified position of the viewer, who is now split and moves among layers of reality, observing how the screen world is being generated. Noteworthily, Douglas sets up his exploration not as an installation environment for the moving viewer, but as classical spectatorship bound to the theatrical setting. The viewer, who is forced to mediate between the different realms, is drawn to participate in the very process of constructing the image, which is not immediately legible, and thus to perform a screening operation of filtering, selecting and reconciling. The viewer emerges as a translucent screen herself partaking in the work of image and narrative fabrication.

The 'throughness' of the translucent screen foregrounds its phenomenological mutability, its power to destabilize space and to trick perception by obscuring the boundaries between surface and space, rendering the very location of the image uncertain. László Moholy-Nagy noted that forms painted on a translucent surface appear to float in the space in front of the pictorial plane, producing a confusion between what is image and what is surrounding environment, which 'bring[s] the surroundings into the picture, while "the surface becomes part of the atmosphere"' (cited in Tóth 2015: 28). Critical responses to Apichatpong Weerasethakul's installation *Dilbar* (2013) allude to such a precariousness: 'The transparent screen had the undesirable effect of making elements not just within the film but actually within the gallery such as the entrance area, the front desk or even the gallery intern part of the film. As such, the transparent screen and the two projectors became a distraction' (Bohr 2014). Not only does everything become a screen surface when touched by projection, but what is more crucial, by the operation of seeing-through, everything 'enters' the image and becomes a part of the screen, appearing absorbed by its diaphanous body. Weerasethakul, for whom drawing the audience's awareness to the screen has always been a component of film and installation practice, manages in his theatrical performance-installation *Fever Room* (2017) to stretch the boundary of the screen as a surface, by going 'beyond conventional flat screens, and creat[ing] distance and a swelling that is traditionally not visible in the space surrounding the screen. As a result, *Fever Room* was able to activate a space that has always been present in movie theaters, have an effect on the audience, and provide some kind of stimulus' (Weerasethakul cited in Sasaki 2017). *Fever Room* literally dissolves screens, replacing them with artificial fog pierced by a beam of light – melting away, the screen becomes the cloud.

Veiling light, luminosity and the atmospheric screen of the Diorama

Expressing his fascination with the Diorama, Anselm Kiefer recognizes it as '"not-yet", as something that must be considered as an ongoing process' – contrary to a painting, which is finalized 'in a solid, coagulated state' (Palais de Tokyo 2017: 276). For Kiefer, the Greek etymology of 'diorama' as 'translucent' suggests a specific throughness, meaning that the image never unfolds on a single level of the surface, but always refers to that which is behind the surface or in the space between layers. Bill Viola's video installation *The Veiling* (1995) stages translucency as passage: light and projection pass through nine successive parallel layers of suspended diaphanous cloth towards the central veil, which becomes an ephemeral locus of the meeting between two apparitions slowly walking towards each other amid a nocturnal forest – two images, a man and a woman, who never coexist in the same video frame but meet fleetingly on the central veil. The projection multiplies and magnifies, while the cloth material diffuses light, and the figures dissipate in intensity and focus as they penetrate further into the scrim layers (Figure 12.2). In Viola's staging of translucency as the relationship between light and matter, the two mediums – the penetrating light and the permeable veil – diffuse each other, making the installation appear enveloped in a haze or transpiring into an atmospheric milieu. According to Viola (1997: 120), the cone of light emerging from each projector 'is articulated in space by the layers of material, revealing its presence as a three-dimensional form that moves through and fills the empty space of the room with its translucent mass'. Zones between layers become activated, air turns visible, and translucency dissolves the boundaries between the surface of the image and the enveloping space. For Viola, as for Kiefer, the image does not merely belong to the single surface but hovers in-between the layers, resulting in a voluminous space of translucency that is dynamically anti-perspectival. Reimagining the screen as a multilayered surface, or rather a place, where the logic of projection meets the logic of layering, *The Veiling* collapses the Cartesian perspective by blurring forms into formlessness, diluting contours in the aerial sfumato. Only the confluence of more than one layer, rather than perspectival depth, produces the effect of the meeting. The screen-as-window, opening towards the homogeneous space secured since the Renaissance by the central perspective, becomes the diaphanous screen-veil, a figure that would correspond to what Thomas Elsaesser (2013: 230) calls 'the new default value of digital vision, presuming a layered, material, yet also mutable and pliable space', a post-pictorial vision substituting the singular frame of linear perspective with multi-perspectivity, layers and veils. When one speaks of an object or environment as being translucent and diffusing light while also being voluminous and thick, one speaks about something as a cloud: the visual

Figure 12.2 *The Veiling*, c. 1995. Bill Viola. Video/sound installation. Two channels of colour video projections from opposite sides of a large dark gallery through nine large scrims suspended from ceiling; two channels of amplified mono sound, four speakers. Scrim size: 8 ft 2 in × 10 ft 10 in (2.4 × 3.3 m) each. Room dimensions: 11 ft 6 in × 24 ft 4 in × 37 ft 10 in (3.5 × 7.4 × 11.5 m). 30 minutes. Performers: Lora Stone, Gary Murphy. Photograph: Roman Mensing.

scenario of *The Veiling* is the passage of light through and inside matter that renders anti-perspectival vision as vision through clouds (see Damisch 2002).

The reciprocity between the translucent veil and the screen was resolved by the nineteenth-century Diorama, which along with the Panorama and the Phantasmagoria – two *dispositifs* distinguished by Casetti (2017) as diverse legacies of the optical screen – was a crucial step towards the invention of the screen in our contemporary sense. I suggest that the Diorama constitutes an additional line of the archaeology of the screen. Like the Panorama and Phantasmagoria, the Diorama screen is both optical and environmental, since the screen is imbricated in the sophisticatedly built environment orchestrating the viewer's passage through layers of space towards the theatre hall with the luminous tableau. Yet it is environmental in an additional sense – the Diorama figures what might be called the atmospheric screen that acts as an extension of three representational paradigms: transparency painting, theatrical curtain and the cloud. Daguerre's Diorama was a dark show of the translucent radiant painting, subjected to carefully orchestrated light to create the illusion of atmospheric movement and transition of times of day. Translucency worked as a crucial constituent for Daguerre's effects – two scenes reproduced on semi-transparent linen canvas, with a landscape or architectural structure painted on both sides, one with translucent oil pigments and another with opaque pigments. When illuminated by light passing from behind the painting or reflected from the painting's frontal surface, one or another image comes to visibility. Passing through the canvas, light induces changes within the pictorial matter itself, allowing Daguerre to succeed in making the still image come alive by one image blending into the next in the course of translucent operation (see Chefranova 2016).

Seen through a vista formed by several diaphanous surfaces concealing the proscenium frame, the dioramic painting produced the effect of a luminous veil stretched across a dark distant background, as if floating in magical ether emitted from the light above and behind. While the Panorama screen shares affinities with the wall, the Diorama screen has recognized the screening nature of the theatrical curtain – one of Daguerre's pictorial skills was painting theatrical curtains with trompe l'œil and light effects (Gernsheim and Gernsheim 1956: 8). Investigating the metaphorical power of the motif of the luminous curtain eclipsing light throughout nineteenth-century aesthetics and painting, Mikhail Iampolsky argues for a genealogy of the Diorama as a part of the corpora of cultural myths and symbolism of luminosity that embraces the allegory of a mystical cover eclipsing truth, the myth of Isis's veil, and the veil covering the blinding light of the sun or the Divine, which all have descended into the vernacular sphere of the theatrical illusionist's attraction and become condensed in a spectacular effect of light penetrating a screen, membrane or medium containing an image (see Yampolsky 1993). The cultural myths still reverberate through *The Veiling*, in which continual images of transformation and light epiphany reflect Viola's preoccupations with a veil effect that aims at imbuing contemporary video art with the ability to approach the sacred. A veil eclipsing a beam of light is described in Edward Bulwer-Lytton's novel *The Last Days of Pompeii* (1842), saturated with multiple evocations of contemporaneous optical media:

> The Egyptian seized the hand of Apecides, and led him wandering … across the chamber towards the curtain at the far end; and now, from behind that curtain, there seemed to burst a thousand sparkling stars; the veil itself, hitherto dark, was now lighted by these fires behind into the tenderest blue of heaven. It represented heaven itself – such a heaven, as in the nights of June might have shone down over the streams of Castaly. Here and there were painted rosy and aerial clouds, from which smiled, by the limner's art, faces of divinest beauty. … And the stars which studded the transparent azure rolled

> rapidly as they shone. … The veil parted – it seemed to vanish into air; and a scene … broke upon the dazzled gaze of the youthful priest. A vast banquet room stretched beyond, blazing with countless lights.
>
> (Bulwer-Lytton 1842: 81)[6]

In the fragment from Lytton, the metaphorical is reduced to the representational; the metaphysics of translucency becomes a pure spectacle. In this description, impacted on by the Diorama's transformation with the melting of images into one another, the curtain turns into a translucent veil that morphs into clouds and the starry sky and then disappears, giving way to spectacular visions of the hall.[7] The surface dissipates, and screening-as-filtering-light becomes included in the family of celestial phenomena and atmospheric media – the starry sky and luminous clouds – suggestive of a homology between the veil and the cloud.

The uncanny effect of the Diorama was grounded in the optical delusion, in what Daguerre termed 'the decomposition of form' – the instability of the visible, the transformations perceived as if happening without any acting force. Images emerge and morph on the translucent screen as if by magic, displaying momentary metamorphoses in mood and atmosphere. The Diorama is infused with atmospheric allusions to ether and fog, retaining among its favourite subjects the volatile and nebulous natural phenomena of mist, clouds and moonlight playing on ruins. One of the first responses to the spectacle states that changes on its screen are analogous to small atmospheric fluctuations; light and shadow transform 'as if clouds were passing over the sun' (Gernsheim and Gernsheim 1956: 14). It is translucency that assists in solving the pictorial problem of rendering 'the aspects of nature … the impressions of diverse changes that are caused, during a given time, by the wind, light, atmosphere, and their modifications' (Daguerre cited in Pinson 2012: 79). In nineteenth-century aesthetics, the dynamic light milieu of cloud as the translucent cover that admits, contains and transmits light becomes an important formal paradigm and metamodel for painting (Yampolsky 1993; also Somaini 2016). John Ruskin (1887: 207), criticizing earlier painters for depicting the world as surfaces and the sky as a flat expanse, notes that painted clouds should be 'filmy', and light should not be *on* clouds but *inside* them. The programme is realized by J. M. W. Turner, who paints the movement of light inside the volume of cloud, fog and sky, representing light emanating from inside the translucent atmospheric layer, while the pictorial canvas itself acts as a veil onto and through which the light can be visualized. This pictorial programme resting upon the cloud shifts the focus from the idea of surface to the idea of atmospheric environment, which is also valid for the Diorama screen that diffuses and emanates light like a cloud. A sketch for décor for the melodrama *Elodie* (Ambigu Comique, 1822), painted at the time when he is working on his first Diorama, shows how Daguerre, renowned for his spectacular stage light effects and illusions, is fascinated by the modulation of sunlight shining through clouds. The invisible source of light enables the canvas to be seen as lit from within, while the very substance of the Diorama's atmospheric mutable screen generates images that are not pre-existing elsewhere and thrown from the outside by projection, but require a coming-through energy of light in order to be perceived. The dioramic images are not *on* the translucent screen but emerging right *inside* the body of the screen, at the moment when light penetrates it. The Diorama's link to the aerial imagination, ruled by equivocation between appearance and disappearance, suggests the phenomenological instability of the translucent screen; suspended in the uncoagulated and unfinalized state that Anselm Kiefer assigns to the Diorama, this luminous screen, perceived as a veil, metamorphosing like a cloud, and transpiring as a real space, mimics matter that defies the surface.

Skin/screen/atmosphere

The series of installations called *Metakimospheres* (2015), created by theoretician and practitioner in the field of performance Johannes Birringer and his team, explores how the body and other performance materials produce what he calls 'kimospheres' or kinetic atmospheres. *Metakimospheres* assembles a constellation of translucent phenomena of skin, screen and atmosphere to engender a wave of relations between body and environment (see Birringer 2016). Several surfaces that Birringer calls 'interskins' are placed inside a darkened gallery studio space: black porous gauzes are stretched around the perimeter of the room, while a white gauze veil is suspended in the centre, slouching down to the floor and draping over several dancers to unify them into a single body. Cameras and sensors, saturating the space, generate particle effects projected from four sides on and through the white gauze (Figure 12.3). Each dancer's movements and breathing directly influence the behaviour of the particles: as the dancers increase the frequency and speed of breathing and movement, the movement of the particles intensifies and multiplies. Textile here functions not as a space partition or architectural medium, but as a malleable translucent envelope for the bodies, one that folds and undulates in tune with corporeal rhythms. Between viscera and atmosphere, the translucent veil inhabits the liminal zone in-between the body and environment, and acts simultaneously as an extension of the outer surface of the body in Marshall McLuhan's (1994: 42, 120) terms; as the second skin for the collective dancer; as a screen for images of breathing; also functioning as a visualization of environment.

The genealogy of kimospheres can be traced to the tradition of the expanded body in dance performance, which inserts textiles in-between the dancer and environment while positioning the body as an extended instrument of space

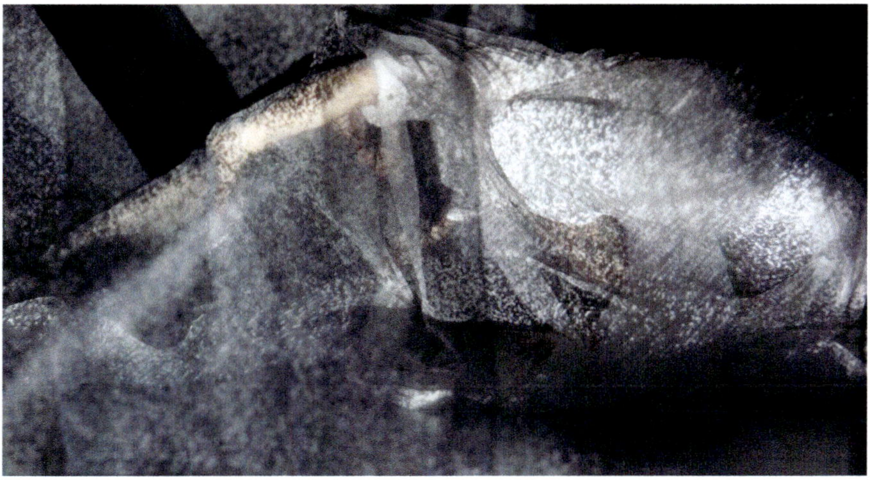

Figure 12.3 *Metakimosphere No.1*, *c.* 2015. Created by DAP-Lab, Artaud Performance Center, Brunel University London. © DAPLab. Photograph: Christopher Bishop.

production by means of the fabric. In his writings on Romantic ballet theatre, Théophile Gautier posits the ballerina, the ethereal creature in diaphanous veils and filmy white skirts, as a modern figure of ephemerality, irrationality and non-Cartesian space. It is modernity's synthesis of contradictions that Gautier (1986: 142) associates with 'a great abuse of white gauze, tulle, and tarlatan', with the white translucent fabric of the ballerina's costume that – functioning materially and metaphorically as a medium filtering everything into one substance – puts environment and bodies into a relation of mimicry. The ballerina and the entire spectacle looming behind the proscenium are products of the technological theatre of the Romantic *ballet blanc*, in which the gauze, along with blue light and artificial fog, constitutes one of its stage technologies. Gautier describes the scenographic environment resulting from new textiles as

> [a] wild landscape in which a hazy moonbeam filtering through the clouds produces a bluish twilight in which the sequins scattered on the ballet skirt of the shade sparkle like dewdrops. Mlle Taglioni *vanishes and materializes again in the mist*, glides over the lake *like a wisp of fog* stirred by the breeze, and displays such fascinating wiles that her lover follows her into the *foam of the waterfall*.
>
> (Gautier 1986: 143, my emphases)

In a technologically fabricated nature, the blue light, saturating space and dematerializing figures make all the elements of the spectacle perceived as if they were made of the same material, or seen through the gauze. With its associations with elemental and atmospheric media, especially water and the misty matter of cloud, the white gauze dissolves the ballerina's body, intangibly merging it with landscape. The diaphanous gauze, linked to the aquatic and aerial imagination, emerges as a site of affective relations between dancer and surrounding environment, while the body functions as the vibrating centre, most vividly presented within the tradition of the expanded performance by Loïe Fuller, whose veils transformed the body into an instrument for making visible the invisible reverberations and waves that suffuse the environment.[8]

Suspending the body between visibility and evanescence, *Metakimospheres*' gauze becomes a screen that allows audiences to contemplate the body *through* its own 'image' – the digital graphics of invisible breathing flowing in and out of the translucent textile. Cocooned inside the nebulous fabric, the body looks as if it is enveloped in its own aura, invisible yet allegedly recorded by early photography in the form of translucent veils and halos. It can be said – following Walter Benjamin, for whom the aura involves veiling – that if the gauzy curtains used as props in early photographic and film studios mimicked the aura, disappearing with mechanical reproduction (see Hansen 2008), *Metakimospheres*' pneumatic screen transpires as another attempt to regain it. Fusing by its etymology 'breath' with 'breeze', the aura is a fleeting waft of air, a subtle emanation, a flicker of atmospheric substance that stands for the inside-out dynamic, for a kind of Platonic essence that manifests itself on the translucent surface and, through its production and kinship with air, exposes the relationship between body, objects and surrounding environment. *Metakimospheres*' screen performs screening as perpetual relays of resonance between atmosphere and bodies that breathe in reflexive correspondence through the translucent and wearable gauze-screen, acting for Birringer as a transmitter and amplifier of the invisible cascade of vibratory effects and reverberations that saturate environments.[9] The translucent screen becomes an instrument by which the body projects itself out and partakes in the technological fabrication of invisible atmosphere, while revealing how contemporary installation art restores aura by assembling an environment and generating new relations.

Postscript

All the metaphors evoked throughout this essay – veil, membrane, skin, atmosphere – return material and phenomenal depth to the screen while underlining its relational epistemic. The aesthetics of translucency secures the screen's post-digital compensatory dream and – through Douglas's post-cinematic memory of indexicality, Viola's peering through the veil only to see more veils, and the body-in-the-veils with the somatic excess of breathing projected on the enveloping membrane – works as a recompense for the superficiality of digital surfaces. With its new materialities, the screen exceeds figuration as window and mirror, but transpires as veil or cloud, and, dissimulating as skin and atmosphere, acts across installation art as a *mise en abyme* of environment. *Metakimospheres*' screen-veil enacts the permeable atmosphere, and Weerasethakul's gesture of swapping the screen for an artificial cloud seems emblematic of a dissolution of the very idea of the screen as surface. Despite their ubiquity, screens seem to melt, and, as Simon Lefebvre (2016: 105) writes in relation to 3D screens, 'disappear into digital territory, they are now the medium used by effects that virtually recreate matter'. The contemporary screen is less a finite object than matter, fluid and adaptable, that does not respect borders and might free the image from the surface to persist between, beyond and through. The utopian gesture to de-localize the image refers back to one of the first employments of the translucent screen in the Phantasmagoria, which created a perceptual conundrum about the location of images: suspended in the air, the projected ghost appeared to be floating in the same phenomenal space with the observer, literally becoming an atmospheric substance. Yet the translucent *dispositif* fundamentally differs from the phantasmagoric *dispositif* – the translucent screen is *both* an ephemeral surface for images *and* the medium through which objects are seen, and, calling attention to itself, it becomes a visible barrier, a lens that may distort the object in the very act of making that object visible. The 'through' turns into 'meta', referring to the oscillation between and beyond opposed poles; such a usage, based on the prefix 'meta' derived from Plato's *metaxis*, was proposed in the definition of the metamodern imaginary as oscillation, characteristic of our era, between aspects of both modernism and postmodernism (Vermeulen and van den Akker 2010) that can equally be attributed to the numerous dualities of contemporary screens. The translucent screen is a metascreen, a chiasmatic surface, implementing a collapse of binaries or in-betweenness as 'modern' principles, in favour of the order of vital simultaneity.

Notes

1 Cathryn Vasseleu (2015: 165) insists on the distinction between transparency and translucency as crucial for digital image synthesis and algorithms that render the appearance of materials with different degrees of translucency, especially human skin.

2 On the material and conceptual mutability of membranes, see Bachner (2016).

3 Exploring how an 'organizational transparency of relations' operates across architectural practices, Eve Blau (2008) suggests that transparency re-emerged as translucency in a MoMA exhibition of *Light Construction* (Terence Riley) in 1995.

4 Looking at a garden pool's ornamental tiles through a layer of water animated by reflections and refractions, Merleau-Ponty (1962: 182) insists: 'I do not see it despite the water and the reflection there; I see it through them and because of them.'

5 Apparatus theory was explored in the work of Jean-Louis Baudry, Jean-Louis Comolli, Jean Narboni, Laura Mulvey and Peter Wollen (see Rosen 1986).

6 I am grateful to Mikhail Iampolsky for drawing my attention to this novel.

7 This transformation resonates with the first reactions to the Diorama as seeing the canvas 'disappearing' and being replaced by a space that seemed real, fostering epistemic uncertainty about whether it was still an image, or a real space opening up through the pictorial veil.

8 On the connections between Loïe Fuller, science and the aesthetics of vibration *c.* 1900, see Iampolski (1996).

9 Birringer adopts the idea of environment as vibrant materialities from Bennett (2010).

List of Figures

List of Tables

Bibliography

Adamson, Glenn (2007), *Thinking Through Craft*, Oxford: Berg.

Adamson, Glenn (2013), *The Invention of Craft*, London: Bloomsbury.

Adamson, Glenn and Victoria Kelley (eds) (2013), *Surface Tensions: Surface, Finish and the Meaning of Objects*, Manchester: Manchester University Press.

Adkins, Lisa and Celia Lury (2009), 'Introduction: What is the Empirical?', *European Journal of Social Theory*, 12 (1): 5–20.

Agamben, Giorgio (2002), 'What is a Paradigm', *European Graduate School Video Lectures*. https://www.youtube.com/watch?v=G9Wxn1L9Er0 (accessed 21 July 2019).

Aldersey-Williams, Hugh (2011), *The Curious Lives of the Elements*, London: Viking/ Penguin.

Allahyari, Morehshin and Daniel Rourke (2016), 'The Additivist Manifesto', in R. Bishop, K. Gensing, J. Parikka and E. Wilk (eds), *Across and Beyond: A Transmediale Reader on Post-digital Practices, Concepts, and Institutions*, 325–33, Berlin: Sternberg.

Allara, David (2015), 'A Perspective on Surfaces and Interfaces', *Nature*, 437 (29): 638–9.

Allen, Stan (2008), *Practice: Architecture, Technique + Representation*, London: Routledge.

Andersen, Hans Christian (1837), *Fairy Tales Told for Children. First Collection. Third Booklet*, Copenhagen: C.A. Reitzel.

Andrew, Sonja (2008), 'Textile Semantics', *Textile*, 6 (1): 32–65.

Anusas, Mike and Cristián Simonetti (eds) (2020), *Surfaces: Transformations of Body, Materials and Earth*, London: Routledge.

Atkinson, Douglas, Pawel Orzechowski, Bruna Petreca, Nadia Bianchi-Berthouze, Penelope Watkins, Sharon Baurley, Stefano Padilla and Mike Chantler (2013), 'Tactile Perceptions of Digital Textiles', *CHI 13 Proceedings of the SIGCHI Conference on Human Factors in Computing Systems*, 1669–78, New York: ACM.

Avezz, Giorgio (2016), 'Intersections between Showing and Concealment in the History of the Concept of Screen', in D. Chateau and J. Moure (eds), *Screens: From Materiality to Spectatorship*, 29–41, Amsterdam: Amsterdam University Press.

Bachner, Andrea (2016), 'Hymenologics: Membrane Politics and the (Un)Making of Difference', in M. Wright and Y. Zhai (eds), *Difference: Sexual, Cultural and Universal*, 14–42, Buffalo, NY: State University of New York at Buffalo.

Balsom, Erika (2013), *Exhibiting Cinema in Contemporary Art*, Amsterdam: Amsterdam University Press.

Barad, Karen (2003), 'Posthumanist Performativity: Toward an Understanding of How Matter Comes to Matter', *Signs: Journal of Women in Culture and Society*, 28 (3): 801–31.

Baronian, Marie-Aude (2017), 'The Dress is the Screen: Dancing Fashion, Dancing Media', *NECSUS*, Autumn.

Barthes, Roland (1990), *S/Z*, London: Jonathan Cape.
Barthes, Roland (1991), 'Rasch', in *The Responsibility of Forms*, 299–312, Berkeley: University of California Press.
Barthes, Roland (1997), 'Le neuter – The Neutral', in *Roland Barthes by Roland Barthes*, 132–3, Berkeley: University of California Press.
Barthes, Roland (2005), *The Neutral*, New York: Columbia University Press.
Battaglia, Andy (2015), '"Helen Lawrence" at BAM: Where Hard-boiled Meets High Tech', *The Wall Street Journal*, 12 October. https://www.wsj.com/articles/helen-lawrence-at-bam-where-hard-boiled-meets-high-tech-1444690935 (accessed 18 February 2019).
Baudrillard, Jean (1988), 'Seduction, or, the Superficial Abyss', in *The Ecstasy of Communication*, 57–75, New York: Semiotext(e).
Baule, Giovanni and Elena Caratti (2017), *Design is Translation*, Milan: Franco Angeli Edizioni.
Benjamin, Walter ([1929] 1978), 'Surrealism', in P. Demetz (ed.), *Reflections: Essays, Aphorisms, Autobiographical Writings*, 177–92, New York: Schocken.
Benjamin, Walter (1999), 'Experience and Poverty', in M. W. Jennings, H. Eiland and G. Smith (eds), *Walter Benjamin: Selected Writings, volume 2, part 2, 1931–1934*, 731–6, Cambridge, MA: Belknap Press.
Benjamin, Walter ([1939] 2003), 'The Work of Art in the Age of its Technological Reproducibility: Third Version', in H. Eiland and M. W. Jennings (eds), *Selected Writings 4: 1938–1940*, 251–83, Cambridge, MA: Harvard University Press.
Benjamin, Walter (2014), 'Erfahrung und Armut', in R. Tiedemann and H. Schweppenhäuser (eds), *Walter Benjamin: Aufsätze, Essays, Vorträge. Gesammelte Schriften, volume II.1*, 213–19, Frankfurt am Main: Suhrkamp.
Bennett, Jane (2010), *Vibrant Matter: A Political Ecology of Things*, Durham, DC: Duke University Press.
Bergson, Henri ([1908] 1988), *Matter and Memory*, New York: Zone Books.
Bergson, Henri (1992), *The Creative Mind: An Introduction to Metaphysics*, New York: Citadel.
Berman, Patricia (2015), 'Dancers on a Plane: Munch, Body, Surface, Projection', *Re-visiting Surface* [conference], 13 November, The National Museum of Art, Oslo.
Berry, David (2015), 'The Postdigital Constellation', in D. Berry and M. Dieter (eds), *Postdigital Aesthetics*, 44–57, Basingstoke: Palgrave Macmillan.
Betts, Paul (2004), *The Authority of Everyday Objects*, Berkeley: University of California Press.
Birringer, Johannes (2016), 'Kimospheres, or Shamans in the Blind Country', *Performance Paradigm*, 12: 7–32.
Bishop, Claire (2012), 'Digital Divide: Contemporary Art and New Media', *Artforum*, September: 434–41.
Blau, Eve (2008), 'Tensions in Transparency. Between Information and Experience: The Dialectical Logic of SANAA's Architecture', *Harvard Design Magazine*, 29: 29–37.
Bletter, Rosemarie Haag (2014), 'Fragments of Utopia: Paul Scheerbart and Bruno Taut', in J. McElheny and C. Burgin (eds), *Glass! Love!! Perpetual Motion!!! A Paul Scheerbart Reader*, 123–9, Chicago: University of Chicago Press.
Bloomfield, Brian P. and Theo Vurdubakis (2005), 'The Secret of Yucca Mountain: Reflections on an Object in Extremis', *Environment and Planning D: Society and Space*, 23: 735–56.
Blowers, Andrew (2017), *The Legacy of Nuclear Power*, London: Routledge.

Bohr, Marco (2014), 'Double Visions by Apichatpong Weerasethakul', *Visual Culture Blog*, 3 June. http://visualcultureblog.com (accessed 18 February 2019).
Bourriaud, Nicolas ([1998] 2002), *Relational Aesthetics*, Dijon: Les presses du réel.
Bower, Stephanie (2016), *The Urban Sketching Handbook: Understanding* Perspective, Beverly, MA: Quarry.
Boyle, Casey (2018), *Rhetoric as a Posthuman Practice*, Columbus: The Ohio State University Press.
Braddock, Sarah and Marie O'Mahoney (2005), *Techno Textiles*, London: Thames and Hudson.
Braidotti, Rosi (2006), *Transpositions: On Nomadic Ethics*, Cambridge: Polity.
Bratton, Benjamin (2016), *The Stack: On Software and Sovereignty*, Cambridge, MA: MIT Press.
Brecht, Bertolt (2013), 'Aus dem Lesebuch für Städtebewohner (1926–27)', in *Bertolt Brecht Hundert Gedichte: Ausgewählt von Siegfried Unseld*, Frankfurt am Main: Suhrkamp.
Brown, Bill (2001), 'Thing Theory', *Critical Inquiry*, 28 (1): 1–22.
Brownell, Blaine Erickson and Becca Casbon (2011), *Matter in the Floating World*, New York: Princeton Architectural Press.
Bruno, Giuliana (2002), *Atlas of Emotion: Journeys in Art, Architecture, and Film*, London; New York: Verso.
Bruno, Giuliana (2007), *Public Intimacy: Architecture and the Visual Arts*, Cambridge, MA: MIT Press.
Bruno, Giuliana (2014), *Surface*: *Matters of Aesthetics, Materiality, and Media*, Chicago and London: University of Chicago Press.
Bruno, Giuliana (2016), 'Storage Space', *e-flux Architecture*. https://www.e-flux.com/architecture/superhumanity/68650/storage-space/ (accessed 3 August 2018).
Bruno, Giuliana (2017), 'Dressing the Surface', *NECSUS*, Autumn: 91–97.
Bruno, Giuliana (forthcoming), *Atmospheres of Projection: Environments of Art and Screen Media*, Chicago: University of Chicago Press.
Bruno, Marco, Simone Carena and Minji Kim (2013), *Borrowed City: Private Use of Public Space in Seoul*, Seoul: Damdi.
Bulwer-Lytton, Edward (1842), *The Last Days of Pompeii*, Leipzig: Bernard Tauchnitz.
Buse, Peter and Andrew Stott (eds) (1999), *Ghosts: Deconstruction, Psychoanalysis, History*, Basingstoke: Palgrave MacMillan.
Byrne, Bryony (n.d.), 'Pristine Realities', *Aesthetica*. http://www.aestheticamagazine.com/pristine-realities/ (accessed 18 February 2019).
Bythell, Duncan (1969), *The Handloom Weavers*, Cambridge: Cambridge University Press.
Campanario, Gabriel (2012), *The Art of Urban Sketching*, Beverly, MA: Quarry.
Canguilhem, Georges (1992), 'Machine and Organism', in J. Crary and S. Kwinter (eds), *Incorporations*, 45–69, New York: Zone.
Carlsen, Emilie (2018), *Emilie Carlsen* [webpage]. http://emiliecarlsen.dk/dopetheme/ (accessed 8 March 2019).
Carpo, Mario (2011), *The Alphabet* and the *Algorithm*, Cambridge, MA: MIT Press.
Casetti, Francesco (2017). 'Notes on a Genealogy of the Excessive Screen', in *Screens* [booklet for Mellon Sawyer Seminar on Genealogies of the Excessive Screen, February–December 2017, Yale University Art Gallery], New Haven, CT: Yale University. https://dev.screens.yale.edu/sites/default/files/files/Screens_Booklet.pdf (accessed 3 September 2018).
Causey, Andrew (2017), *Drawn to See: Drawing as an Ethnographic Method*, Toronto: University of Toronto Press.

Cavell, Stanley (1971), *The World Viewed: Reflections on the Ontology of Film*, Cambridge, MA: Harvard University Press.
Chateau, Dominique and José Moure (eds) (2016), *Screens: From Materiality to Spectatorship*, Amsterdam: Amsterdam University Press.
Chefranova, Oksana (2016), 'Promenade through the Theater of Illusion: *Dioramas* in Palais de Tokyo', *NECSUS*, Autumn: 217–232.
Cheng, Anne Anlin (2011), *Second Skin: Josephine Baker and the Modern Surface*, New York: Oxford University Press.
Clark, Andy (2003), *Natural-Born Cyborgs: Minds, Technologies, and the Future of Human Intelligence*, Oxford: Oxford University Press.
Clynes, Manfred E. and Nathan S. Kline (1960), 'Cyborgs and Space', *Astronautics*, September: 26–7; 74–6.
Coblentz, Cassandra (ed.) (2003), *Surface Tension* [exhibition catalogue], Philadelphia: The Fabric Workshop and Museum.
Coleman, Rebecca and Liz Oakley-Brown (2017), 'Visualizing Surfaces, Surfacing Vision: Introduction', *Theory, Culture & Society*, 34 (7–8): 5–27.
Conley, Tom (2006), *Cartographic Cinema*, Minneapolis, MN: University of Minnesota Press.
Conlon, John J. (1980), 'Walter Pater's "Diaphaneite"', *English Language Notes*, XVII (3): 195.
Coole, Diana and Samantha Frost (eds) (2010), *New Materialisms: Ontology, Agency, and Politics*, Durham, NC: Duke University Press.
Corning Museum of Glass (1999), *Innovations in Glass*, New York: Corning Museum of Glass.
Coulthard, Lisa (2008), 'Uncanny Memories: Stan Douglas, Subjectivity and Cinema', *Scope: An Online Journal of Film Studies*, 12: 1–14.
Crafts Council (1986), *David Pye: Wood Carver and Turner*, London: Crafts Council.
Crafts Council (2020) Crafts Council. http://craftscouncil.org.uk (accessed 22 July 2020).
Crafts Council (n.d.), *Curved Dish (W43)*. http://collections.craftscouncil.org.uk/object-w43 (accessed 1 February 2019).
Cramer, Florian (2015), 'What is "Post-digital"?' in D. Berry and M. Dieter (eds), *Postdigital Aesthetics*, 12–26, Basingstoke: Palgrave Macmillan.
Crary, Jonathan and Sanford Kwinter (eds) (1992), *Incorporations*, New York: Zone.
Crawford, Matthew (2010), *The Case for Working with your Hands, or, Why Office Work is Bad for Us and Fixing Things Feels Good*, London: Penguin.
Cullen, Gordon (1961), *The Concise Townscape*, Architectural Press.
D'Aloia, Adriano, Marie-Aude Baronian and Marco Pedroni (2017), 'Fashionating Images. Audiovisual Media Studies Meet Fashion', *Comunicazioni sociali*, 1: 3–13.
Damisch, Hubert (2002), *A Theory of Cloud: Toward a History of Painting*, Stanford, CA: Stanford University Press.
Darley, Andrew (2002), *Visual Digital Culture: Surface Play and Spectacle in New Media Genres*, 58–77, London: Routledge.
Davidson, Rosemary and Arzu Tahsin (2018), *Craftfulness*, London: Quercus.
Dawson, Ian (2015) 'Press Release', *The Wanderer's Nightsong II*, C&C Gallery, London, 3 April–10 May 2015 [exhibition]. http://www.iandawsonstudio.com/the-wanderer-s-nightsong-ii.html (accessed 30 July 2018).
DeLanda, Manuel (1997) 'The Machinic Phylum', *V2_Lab for the Unstable Media*. https://v2.nl/archive/articles/the-machinic-phylum (accessed 29 August 2017).
DeLanda, Manuel (1998), 'Deleuze and the Genesis of Form', *Art Orbit*, 1 (1): 1–6.
DeLanda, Manuel (2012), 'Interview with Manuel DeLanda', in R. Dolphijn and I. van der Tuin (eds), *New Materialism: Interviews and Cartographies*, 38–47, Ann Arbor, MI: Open Humanities Press.

DeLanda, Manuel (2016), *Assemblage Theory*, Edinburgh: Edinburgh University Press.
Deleuze, Gilles (1988), *Foucault*, Minneapolis, MN: University of Minnesota Press.
Deleuze, Gilles (1993), *The Fold: Leibniz and the Baroque*, Minneapolis, MN: University of Minnesota Press.
Deleuze, Gilles (1994), *Difference and Repetition*, New York: Columbia University Press.
Deleuze, Gilles and Félix Guattari (1986), *Kafka: Toward a Minor Literature*, Minneapolis, MN: University of Minnesota Press.
Deleuze, Gilles and Félix Guattari (1987), *A Thousand Plateaus: Capitalism and Schizophrenia*, Minneapolis, MN: University of Minnesota Press.
Deleuze, Gilles and Félix Guattari (2009), 'Balance-Sheet Program for "Desiring Machines"', in S. Lotringer (ed.), *Chaosophy: Texts and Interviews 1972–1977*, 90–115, Los Angeles: Semiotext(e).
Deleuze, Gilles and Claire Parnet (2007), *Dialogues II*, New York: Columbia University Press.
Della Dora, Veronica (2015), 'The Mantle of the Earth: Surfaces, Landscapes and Aesthetics', in H. Hawkins and E. Straughan (eds), *Geographical Aesthetics*, 225–50, London; New York: Routledge.
DeNichola, Lane (2016), 'Forging Source: Considering the Craft of Computer Programming', in C. M. Wilkinson-Weber and A. O. DeNicola (eds), *Critical Craft*, 35–55, London: Bloomsbury.
Denson, Shane and Julia Leyda (eds) (2016), *Post-cinema: Theorizing 21st-century Film*, Falmer: Reframe.
Denzin, Norman and Yvonna Lincoln (eds) (2018), *The SAGE Handbook of Qualitative Research*, 5th edn, London: SAGE.
Derrida, Jacques (2005), *Paper Machine*, Stanford, CA: Stanford University Press.
De Ruysser, Tine (2009), 'Wearable Metal Origami? The Design and Manufacture of Metallised Folding Textiles', PhD thesis, Royal College of Art, London.
de Wolfe, Ivor (2013), *The Italian Townscape*, London: Artifice.
Doane, Mary Ann (2007), 'The Indexical and the Concept of Medium Specificity', *differences*, 18 (1): 128–52.
Doane, Mary Ann (2012), 'Notes from the Field: Contingency', *Art Bulletin*, 94 (3): 344–61.
Dolphijn, Rick and Iris van der Tuin (eds) (2012), *New Materialism: Interviews and Cartographies*, Ann Arbor, MI: Open Humanities Press.
Dormer, Peter (1997), 'David Pye', in M. Coatts (ed.) *Pioneers of Modern Craft*, 71–82, Manchester: Manchester University Press.
Eadie, Leslie and Tushar K. Ghosh (2011), 'Biomimicry in Textiles', *Journal of the Royal Society Interface*, 8 (59): 761–75.
Easterling, Keller (2014), *Extrastatecraft: The Power of Infrastructure Space*, London: Verso.
Ellsworth, Elizabeth and Jamie Kruse (eds) (2013), *Making the Geologic Now*, New York: Punctum.
Elsaesser, Thomas (2013), 'The "Return" of 3-D: On Some of the Logics and Genealogies of the Image in the Twenty-First Century', *Critical Inquiry*, 39 (2): 217–46.
Epstein, Jean (2012), 'The Fluid World of the Screen', in S. Keller and J. N. Paul (eds), *Jean Epstein: Critical Essays and New Translations*, Amsterdam: Amsterdam University Press.
Ernst, Wolfgang (2015), 'Between the Archive and the Anarchivable', *Mnemoscape*, 1.
Evans, Caroline (2013), *The Mechanical Smile: Modernism and the First Fashion Shows in France and America, 1900–1929*, New Haven, CT: Yale University Press.

Evans, Robin (1996), 'The Developed Surface: An Enquiry into the Brief Life of an Eighteenth-century Drawing Technique', in R. Evans (ed.), *Translations from Drawing to Building and Other Essays*, 195–232, London: Architectural Association.

Fast, Heidi, Taru Leppänen and Milla Tiainen (2018), 'Vibration', *New Materialism* [webpage], 30 March. https://newmaterialism.eu/almanac/v/vibration.html (accessed 10 March 2019).

Fletcher, Kate and Mathilda Tham (eds) (2014), *The Routledge Handbook of Sustainability and Fashion*, London; New York: Routledge.

Flusser, Vilém (1991), 'Digitaler Schein', in F. Roetzer (ed.), *Aesthetik der Electronischen Medien*, Frankfurt am Main: Fischer Verlag.

Flusser, Vilém (1999), *The Shape of Things: A Philosophy of Design*, London: Reaktion.

Flusser, Vilém (2000), *Towards a Philosophy of Photography*, London: Reaktion.

Flusser, Vilém ([1985] 2011), *Into the Universe of Technical Images*, Minneapolis, MN: University of Minnesota Press.

Flusser, Vilém (2011), *Does Writing Have a Future?* Minneapolis, MN: University of Minnesota Press.

Flusser, Vilém and Louis Bec ([1987] 2012), *Vampyroteuthis Infernalis*, Minneapolis, MN: University of Minnesota Press.

Forsyth, Isla, Hayden Lorimer, Peter Merriman and James Robinson (2013), 'What Are Surfaces?' *Environment and Planning A*, 45 (5): 1013–20.

Foster, Hal (1993), *Compulsive Beauty*, Cambridge, MA: MIT Press.

Foucault, Michel (2002), *The Archaeology of Knowledge*, London: Routledge.

Frayling, Christopher (2011), *On Craftsmanship: Towards a New Bauhaus*, London: Oberon.

Fremont, Diane (2017), 'The Spiracle in Alchemy and Art', *ARAS Connections*, 3. https://aras.org/sites/default/files/docs/00119Fremont.pdf (accessed 3 January 2019).

Friedberg, Anne (2006), *The Virtual Window: From Alberti to Microsoft*, Cambridge, MA: MIT Press.

Gaboury, Jacob (2015), 'Hidden Surface Problems: On the Digital Image as Material Object', *Journal of Visual Culture*, 14 (1): 40–60.

Galpin, Richard (1998), 'Erasure in Art: Destruction, Deconstruction, and Palimpsest', *Richard Galpin Studio*. https://www.richardgalpin.co.uk/erasureinart (accessed 19 February 2020).

Gandelsonas, Mario (1999), *X-Urbanism: Architecture and the American City*, New York: Princeton Architectural Press.

Gautier, Théophile (1986), 'Opera: Marie Taglioni's Benefit', in *Gautier on Dance*, London: Dance Books.

Geesin, Frances (1995), 'The Chemical and Structural Manipulation of Fabrics and Fibres through Stiffening Techniques with Specific Emphasis on Electrodeposition', PhD Thesis, Royal College of Art, London.

Gell, Alfred (1998), *Art and Agency: An Anthropological Theory*, Oxford: Clarendon Press.

Gell, Alfred ([1985] 2013), 'The Network of Standard Stoppages', in L. Chua and M. Elliot (eds), *Distributed Objects: Meaning and Mattering after Alfred Gell*, 88–113, New York: Berghahn.

Gernsheim, Helmut and Alison Gernsheim (1956), *L. J. M. Daguerre (1787–1851), The World's First Photographer*, Cleveland: World Pub.

Gibson, James ([1979] 1986), *The Ecological Approach to Visual Perception*, New York: Psychology Press.

Gitelman, Lisa (2014), *Paper Knowledge: Towards a Media History of Documents*, Durham, NC: Duke University Press.
Goldsmith, Kenneth (2011), *Uncreative Writing: Managing Language in the Digital Age*, New York: Columbia University Press.
Goldsmith, Kenneth (2012), 'Conceptual Writing: A Worldview', *Poetry Foundation*. http://www.poetryfoundation.org/harriet/2012/04/conceptual-writing-a-worldview/ (accessed 1 August 2018).
Gomel, Elana and Stephen A. Weninger (2004), 'Romancing the Crystal: Utopias of Transparency and Dreams of Pain', *Utopian Studies*, 15 (2): 65–91.
Goodman, Nelson (1976), *Languages of Art*, Indianapolis: Hackett.
Gorfinkel, Elena (2018), 'Promiscuous Histories, Materialist Theories, Speculative Poetics', *Cinema Journal*, 57 (2): 121–5.
Graves, Jane (2002), 'Symbol, Pattern and the Unconscious', in M. Schoeser and C. Boydell (eds), *Disentangling Textiles*, 45–55, London: Middlesex University Press.
Gregson, Nicky (2012), 'Projected Futures: The Political Matter of UK Higher Activity Radioactive Waste', *Environment and Planning A*, 44: 2006–22.
Grimshaw, David (2017), 'Crafting the Digital', *The Design Journal*, 20 (sup1): s3735–s3748.
Grimshaw, David (n.d.), 'Homage to David Pye', *3DDesign: Manchester School of Art*. https://3ddesignmanchesterschoolofart.wordpress.com/2015/04/29/digital-making-a-future-for-materiality-in-a-digital-world/ (accessed 29 August 2018).
Grombacher, Pamela (2017), 'What is Digital Materiality?' *Rethinking Matter*, 27 March. http://www.rethinkingmatter.com/blog/2017/3/27/what-is-digital-materiality (accessed 8 March 2019).
Gropius, Walter ([1937] 1989), *The New Architecture and the Bauhaus*, Cambridge, MA: MIT Press.
Grosz, Elizabeth (2008), *Chaos, Territory, Art: Deleuze and the Framing of the Earth*, New York: Columbia University Press.
Guattari, Félix ([1992] 1995), *Chaosmosis: An Ethico-aesthetic Paradigm*, Bloomington; Indianapolis: Indiana University Press.
Guillerme, Jacques, Hélène Vérin and Stephen Sartarelli (1989), 'The Archaeology of Section', *Perspecta*, 25: 226–57.
Guy, John (2013), '"One Thing Leads to Another": Indian Textiles and the Early Globalization of Style', in A. Peck (ed.), *Interwoven Globe: The Worldwide Textile Trade, 1500–1800*, 12–27, New York: Metropolitan Museum of Art.
Hacking, Ian (1998), 'Canguilhem amid the Cyborgs', *Economy and Society*, 27 (2–3): 202–16.
Haggerty, Kevin and Richard Ericson (2000), 'The Surveillant Assemblage', *British Journal of Sociology*, 51 (4): 605–22.
Hallam, Elizabeth and Tim Ingold (eds) (2014), *Making and Growing: Anthropological Studies of Organisms and Artefacts*, London: Routledge.
Halprin, Lawrence (1966), *Freeways*, New York: Van Nostrand Reinhold.
Halprin, Lawrence (1982), *R.S.V.P. Cycles*, New York: George Braziller.
Hansen, Miriam Bratu (2008), 'Benjamin's Aura', *Critical Inquiry*, 34 (2): 336–75.
Hansmeyer, Michael (2013), 'The Digital Grotesque I, with Benjamin Dillenburger', *Michael Hansmeyer* [webpage]. http://www.michael-hansmeyer.com/digital-grotesque-I (accessed 21 July 2019).
Haraway, Donna (1991), 'A Cyborg Manifesto', in *Simians, Cyborgs and Women: The Reinvention of Nature*, 149–81, New York: Routledge.

Hardcastle, Lucy (2019), 'Archetypes', *Lucy Hardcastle* [webpage]. https://www.lucyhardcastle.com/archetypes (accessed 5 April 2019).
Harrod, Tanya (ed.) (2018), *Craft*, Cambridge, MA: MIT Press.
Harvey, Penny (2019), 'Lithic Vitality. Human Entanglement with Non-organic Matter', in P. Harvey, C. Krohn-Hansen and K. G. Nustad (eds), *Anthropos and the Material*, 184–208, Durham, NC: Duke University Press.
Hauser, Jens (ed.) (2008), *Sk-interfaces*, Liverpool: Liverpool University Press.
Hecht, Gabrielle (2018), 'Interscalar Vehicles for an African Anthropocene: On Waste, Temporality, and Violence', *Cultural Anthropology*, 33 (1): 109–41.
Hedges, Susan, Tina Engels-Schwarzpaul and Ross Jenner (2017), 'Surface/Pattern: A Pursuit of Material Narratives', *Interstices*, 18: 2–10.
Heidegger, Martin (1954), 'The Thing', in *Poetry, Language, Thought*, 163–80, New York: Perennial Library.
Helmer, Magdelena (2005), 'Surfaces and Interfaces', Nature, 437: 637. https://doi.org/10.1038/437637a.
Hoskins, Tansy E. (2014), *Stitched Up: The Anti-capitalist Book of Fashion*, London: Pluto.
Huhtamo, Erkki (2004), 'Elements of Screenology: Toward an Archaeology of the Screen', *Iconics: International Studies of the Modern Image*, 7: 31–82.
Hui, Yuk (2015), 'Towards a Relational Materialism', *Digital Culture and Society*, 1 (1): 131–47.
Hui, Yuk and Andreas Broeckmann (eds) (2015), 'Introduction', in *30 Years after Les Immatériaux*, 9–24, Lüneburg: Meson.
Iampolski, Mikhail (1996), 'La danse, le cinéma et un peu de physique', *Le Cinematheque*, 9: 95–103.
Ialenti, Vincent F. (2014), 'Adjudicating Deep Time: Revisiting the United States' High-level Nuclear Waste Repository Project at Yucca Mountain', *Science & Technology Studies*, 27 (2): 27–48.
Igoe, Elaine (2013), 'In Textasis: Matrixial Narratives of Textile Design', PhD thesis, Royal College of Art, London.
Iles, Chrissie (2007), 'Surface Tension', in Francesco Bonami (ed.), *Rudolf Stingel* [exhibition catalogue], 23–9, New Haven, CT: Yale University Press.
Imperiale, Alicia (2000), *New Flatness: Surface Tension in Digital Architecture*, Basel: Birkhäuser.
Ingold, Tim (2000), *The Perception of the Environment*, London: Routledge.
Ingold, Tim (2007a), *Lines: A Brief History*, London: Routledge.
Ingold, Tim (2007b), 'Materials Against Materiality', *Archaeological Dialogues*, 14 (1): 1–16.
Ingold, Tim (2010), 'The Textility of Making', *Cambridge Journal of Economics*, 34: 91–102.
Ingold, Tim (2011), *Being Alive*, London and New York: Routledge.
Ingold, Tim (2013), *Making*, Abingdon: Routledge.
Ingold, Tim (2017), 'Surface Visions', *Theory, Culture & Society*, 34 (7–8): 99–108.
International Council on Archives (2000), *ISAD(G) (2000) General International Standard Archival Description*, 2nd edn, Ottawa: International Council on Archives.
Irvine, Richard D. G. (2014), 'Deep Time: An Anthropological Problem', *Social Anthropology/Anthropologie Sociale*, 22 (2): 157–72.
Jackson, Michael (2012), *Lifeworlds: Essays in Existential Anthropology*, Chicago: University of Chicago Press.

Jameson, Fredric (1990), *Postmodernism, or, the Cultural Logic of Late Capitalism*, Durham, NC: Duke University Press.
Jiang, Kinor (2009), *Fabric: Metallic Textile Design*, Hong Kong: Sandhu.
Johnston, Ian (2017), 'AI Robots Learning Racism, Sexism and Other Prejudices from Humans, Study Finds', *The Independent*, 13 April. https://www.independent.co.uk/life-style/gadgets-and-tech/news/ai-robots-artificial-intelligence-racism-sexism-prejudice-bias-language-learn-from-humans-a7683161.html (accessed 25 March 2020).
Johnston, Lucy (2015), *Digital Handmade: Craftsmanship in the New Industrial Revolution*, London: Thames and Hudson.
Jones, Caroline A. (ed.) (2006), *Sensorium: Embodied Experience, Technology, and Contemporary Art*, Cambridge, MA: MIT Press.
Jones, Kathryn (2012), 'Victoria & Albert, Art & Love', London: Royal Collection Trust. https://www.rct.uk/sites/default/files/V%20and%20A%20Art%20and%20Love%20%28Jones%29.pdf (accessed 20 August 2019).
Joselit, David (2006), 'Surface Vision', in N. Baume (ed.), *Super Vision* [exhibition catalogue], Cambridge, MA: MIT Press.
Jung, Carl (1970), *Collected Works of C. G. Jung, 14: Mysterium Coniunctionis*, G. Adler and R. F. C. Hull (eds), Princeton: Princeton University Press.
Kalshoven, Petra Tjitske (2020), 'An Anthropological Perspective on the Siting Process for a GDF in England: Values and Worldviews amongst Stakeholders in West Cumbria' [conference], *Nuclear Energy Agency IGSC Safety Case Symposium 2018*.
Kaltenbach, Frank (ed.) (2004), *Translucent Materials: Glass, Plastics, Metals*, Birkhäuser: Edition Detail.
Kara, Selmin (2016), 'Anthropocenema: Cinema in the Age of Mass Extinctions', in S. Denson and J. Leyda (eds), *Post-cinema: Theorizing 21st-century Film*, 750–84, Falmer: Reframe.
Keith, Sara (2010), 'Silver as Dye: The Organic Idiom of Shibori and Electro Deposition', PhD thesis, University of Dundee.
Kepes, György (1944), *The Language of Vision*, Chicago: Paul Theobald.
Kettle, Alice, Victoria Brown, David Grimshaw, Kate Egan and Fabrizio Cocchiarella (2014), *A Place at the Table* (Pallant House, Chichester, 25 November 2014–18, January 2015) [exhibition]. http://e-space.mmu.ac.uk/id/eprint/756 (accessed 29 August 2018).
Kinchin, Juliet (2012), 'The Crystal Chain and Architectural Play', in J. Kinchin and A. O'Connor (eds), *Century of the Child: Growing by Design, 1900–2000*, 60–2, New York: MoMA.
Korn, Peter ([2013] 2017), *Why We Make Things & Why It Matters: The Education of a Craftsman*, London: Vintage.
Koutsoumpos, Leonidas (2018), 'Drawing Sections', *Art, Materiality and Representation* (Royal Anthropological Institute) [conference].
Kracauer, Siegfried (1987), 'Cult of Distraction. On Berlin's Picture Palaces', *New German Critique*, 40 (Winter): 91–6.
Kracauer, Siegfried (1995), *The Mass Ornament: Weimar Essays*, Cambridge, MA: Harvard University Press.
Küchler, Susanne (2008), 'Technological Materiality: Beyond the Dualist Paradigm', *Theory, Culture & Society*, 25 (1): 101–20.
Lambourne, Lionel (1986), 'Artistic Affinities', in *David Pye: Wood Carver and Turner*, 21–4, London: Crafts Council.

Laughlin, Zoe (2010), 'Beyond the Swatch: How Can the Science of Materials Be Represented by the Materials Themselves in a Materials Library?', PhD thesis, King's College London.

Lawrie, Eleanor (2019), 'Could a Computer Ever Create Better Art than a Human?', *BBC News*, 23 April. https://bbc.in/2pj4TLl (accessed 23 April 2019).

Leatherbarrow, David and Mohsen Mostafavi (2002), *Surface Architecture*, Cambridge, MA: MIT Press.

Lefebvre, Simon (2016), 'The Disappearance of the Surface', in D. Chateau and J. Moure (ed.), *Screens: From Materiality to Spectatorship*, Amsterdam: Amsterdam University Press.

Leonard, Zeke (2011), 'Tools', *Shavings and Dust* [webpage], 6 April. http://shavingsanddust.blogspot.com/2011/04/tools.html (accessed 18 December 2018).

Leonardi, Paul M. (2010), 'Digital Materiality?', *First Monday*, 15 (6). https://firstmonday.org/article/view/3036/2567 (accessed 8 March 2019).

Leslie, Esther (2013), 'Dreams for Sale', in M. Uhlirova (ed.), *Birds of Paradise: Costume as Cinematic Spectacle*, 29–42, London: Koenig.

Lewis, Paul, Marc Tsurumaki and David J. Lewis (2016), *The Manual of Section*, New York: Princeton Architectural Press.

Loos, Adolf (2002), 'The Story of the Poor Little Rich Man (1900)', in *Adolf Loos: On Architecture*, 48–52, Riverside, CA: Ariadne Press.

Lotringer, Sylvère, and Paul Virilio (2005), *The Accident of Art*, Cambridge, MA: Semiotext(e).

Lottersberger, Anna (2012), 'Design, Innovation and Competitiveness in the Textile Industry: Upstream Design Innovation', PhD thesis, Politecnico Di Milano.

Lucas, Ray (2009a), 'The Sensory Experience of Sacred Space: Senso-Ji and Meiji-Jingu, Tokyo', *MONU: Magazine on Urbanism*, 10: 46–55.

Lucas, Ray (2009b), 'Designing Ambiances: Vocal Ikebana and Sensory Notation', in *Creating an Atmosphere Proceedings 2008*, Grenoble: CRESSON. https://halshs.archives-ouvertes.fr/halshs-00833931/document (accessed 9 August 2019).

Lucas, Ray (2009c), 'Designing a Notation for the Senses', in *Architectural Theory Review*, 14 (2): 173–92.

Lucas, Ray (2016), 'Inscriptive Practice as Gesture', in L. Pearson et al. (eds), *Drawing Futures*, 217–23, London: UCL Press.

Lucas, Ray (2017a), 'Electricity as Limit in Namdaemun Market, Seoul', in J. Cross et al. (eds), *Our Lives with Electric Things*. https://culanth.org/fieldsights/series/our-lives-with-electric-things (accessed 9 August 2019).

Lucas, Ray (2017b), 'Castors: Facilitators of Impermanent and Mobile Architecture', in R. Harkness (ed.), *An Unfinished Compendium of Materials*. https://knowingfromtheinside.org/files/#about (accessed 9 August 2019).

Lucas, Ray (2017c), 'Tarpaulin: Denoting Opening and Closure', in R. Harkness (ed.), *An Unfinished Compendium of Materials*. https://knowingfromtheinside.org/files/#about (accessed 9 August 2019).

Lucas, Ray (2017d), 'Bristol Board: A Self-effacing Drawing Technology', in R. Harkness (ed.), *An Unfinished Compendium of Materials*. https://knowingfromtheinside.org/files/#about (accessed 9 August 2019).

Lucas, Ray (2017e), 'The Discipline of Tracing in Architectural Drawing', in C. Johannessen and T. Van Leeuwen (eds), *The Materiality of Writing*, 116–37, London: Routledge.

Lucas, Ray (2018), 'Script and Score: Revisiting Nelson Goodman at Sanja Matsuri', in J. Browne, C. Frost and R. Lucas (eds), *Architecture, Festival and the City*, 81–96, London: Routledge.
Lucas, Ray (2019), *Drawing Parallels: Knowledge Production in Axonometric, Isometric, and Oblique Drawings*, London: Routledge.
Lucas, Ray (2021), 'Namdaemun Market (Seoul)', in M. Vellinga (ed.), *Encyclopaedia of Vernacular Architecture of the World*, London: Bloomsbury.
Lucas, Ray (2020a), *Anthropology for Architects: Social Relations and the Built Environment*, London: Bloomsbury.
Lucas, Ray (2020b), 'Threshold as a Social Surface: The Architecture of South Korean Urban Marketplaces', in C. Simonetti and M. Anusas (eds), *Surfaces: Transformations of Body, Materials and Earth*, 97–115, London: Routledge.
Lucas, Ray and Ombretta Romice (2008), 'Representing Sensory Experience in Urban Design', *Design Principles and Practices*, 2 (4): 83–94.
Lucas, Ray and Ombretta Romice (2010), 'Assessing the Multi-sensory Qualities of Urban Space', *Psyecology*, 1 (2): 263–76.
Ludovico, Alessandro (2012), *Post-digital Print: The Mutation of Publishing since 1894*, Eindhoven: Onomatopee.
Lynch, Kevin (1964), *The Image of the City*, Cambridge, MA: MIT Press.
MacArthur, John (2007), *The Picturesque: Architecture, Disgust and Other Irregularities*, London: Routledge.
Machado, Pedro (2009), 'Awash in the Sea of Cloth: Gujarat, Africa, and the Western Indian Ocean, 1300–1800', in G. Riello and P. Parthasarathi (eds), *The Spinning World: A Global History of Cotton Textiles, 1200–1850*, 161–79, Oxford: Oxford University Press.
Mackey, Angella (2019), *PHEM* [webpage]. https://phem.design/what-is-phem (accessed 5 February 2019).
Mall, Amar (2007), 'Structure, Innovation and Agency in Pattern Construction: The Kōlam of Southern India', in E. Hallam and T. Ingold (eds), *Creativity and Cultural Improvisation*, 55–78, Oxford: Berg.
Manaugh, Geoff and Nicola Twilley (2013), 'One Million Years of isolation: An Interview with Abraham van Luik', in E. Ellsworth and J. Kruse (eds), *Making the Geologic Now*, 226–37, New York: Punctum.
Mandelup, Liza (2016), 'The Fifth Sense, Episode 3: Making Codes with Lucy Hardcastle', *i-D*, 13 December. https://www.lucyhardcastle.com/about (accessed 8 March 2019).
Mann, Thomas ([1924] 1995), *The Magic Mountain*, New York: Vintage International.
Manske, Beate (2012), 'Wilhelm Wagenfelds Rautenglas—Aufbau und Vermarktung des Sortiments', in *Zeitgemäß und Zeitbeständig. Industrieformen von Wilhelm Wagenfeld*, 148–185, Bremen: Wilhelm Wagenfeld Stiftung.
Manzini, Ezio (1989), *The Material of Invention*, Cambridge, MA: MIT Press.
Marks, Laura (2000), *Skin of the Film*, Durham, NC: Duke University Press.
Mattern, Shannon (2016), 'Cloud and Field', *Places* (August). https://placesjournal.org/article/cloud-and-field/ (accessed 16 July 2018).
Matthewson, Gill (2009), 'People Who Live in Glass Houses: Walter Benjamin and the Dream of Glass Architecture', in J. Gatley (ed.), *Cultural Crossroads: Proceedings of the 26th Annual Conference of the Society of Architectural Historians*, New Zealand, Auckland: SAHANZ.

Maturana, Humberto and Francisco Varela (1980), 'Autopoiesis: The Organization of the Living', 73–123, in *Autopoiesis and Cognition*, Dordrecht: D. Reidel.
Mauss, Marcel ([1935] 2006), 'Techniques of the Body', in N. Schlanger (ed.), *Techniques, Technology and Civilisation*, 77–96, Oxford: Berghahn.
McCullough, Malcolm (1996), *Abstracting Craft: The Practiced Digital Hand*, Cambridge, MA: MIT Press.
McLuhan, Marshall (1994), *Understanding Media: The Extensions of Man*, Cambridge, MA: MIT Press.
McQuaid, Matilda (1998), 'Transformation: Process and Techniques in Contemporary Japanese Textiles', in C. McCarty and M. McQuaid (eds), *Structure and Surface*, 19, New York: MoMA.
Merleau-Ponty, Maurice (1962), *The Phenomenology of Perception*, London: Routledge.
Mertins, Detlef (1996), 'The Enticing and Threatening Face of Prehistory: Walter Benjamin and the Utopia of Glass', *Assemblage*, 29 (April): 6–23.
Miller, Daniel (2005), 'Style and Ontology', in J. Friedman (ed.), *Consumption and Identity*, 53–70, London: Routledge.
Miller Lane, Barbara (1968), *Architecture and Politics in Germany 1918–1945*, Cambridge, MA: Harvard University Press.
Mitchell, Victoria (2012), 'Textiles, Text, Techne', in J. Hemmings (ed.), *The Textile Reader*, 5–13, Oxford: Berg.
Mitchell, W. J. T. (2015), 'Screening Nature (and the Nature of the Screen)', *New Review of Film and Television Studies*, 13 (3): 231–46.
Morgan, Daniel (2011), 'The Afterlife of Superimposition', in D. Andrew (ed.), *Opening Bazin: Postwar Film Theory and Its Afterlife*, 127–41, Oxford: Oxford University Press.
Mulvey, Laura (2006), *Death 24x a Second: Stillness and Moving Image*, London: Reaktion.
Mühleck, Georg (2019), 'Caspecies [2014–2015]', *Georg Mühleck Archive* [webpage]. http://www.georgmuehleck.com/artworks/caspecies-2014-2015/ (accessed 30 July 2019).
Murray, Scott (2010), *Translucent Building Skins: Material Innovations in Modern and Contemporary Architecture*, London: Routledge.
Neumann, Dietrich (2013), 'The House that Mies Built: A Luminous Architectural Gem Becomes a Black Box for Political Fantasy', *The Berlin Journal*, 25 (Fall): 12–15.
Nunberg, Geoffrey (1996), 'Farewell to the Information Age', in G. Nunberg (ed.), *The Future of the Book*, 1–23, Berkeley, CA: University of California Press.
O'Brien, Hettie (2019), 'Ghosts in the Machines: The Invisible Human Labour behind AI', *New Statesman*, 5 June. https://www.newstatesman.com/2019/06/ghosts-machines-invisible-human-labour-behind-ai (accessed 8 June 2019).
Openshaw, Jonathan (ed.) (2015), *Postdigital Artisans: Craftsmanship with a New Aesthetic in Fashion, Art, Design and Architecture*, London: Frame.
Oppenheimer, Sarah (2014), 'Giuliana Bruno by Sarah Oppenheimer [interview]', *Bomb Magazine*, July. https://bombmagazine.org/articles/giuliana-bruno/ (accessed 4 August 2018).
Palais de Tokyo (2017), *DIORAMAS*, Paris: Flammarion.
Parikka, Jussi (2012), *What is Media Archaeology?* Cambridge: Polity.
Parikka, Jussi (2015), *A Geology of Media*, Minneapolis, MN: University of Minnesota Press.
Parissien, Steven (2014), *The Life of the Automobile: The Complete History of the Motor-Car*, Thomas Dunne Books.

Parks, Lisa and Nicole Starosielski (eds) (2015), *Signal Traffic: Critical Studies of Media Infrastructures*, Urbana, IL: University of Illinois Press.
Parr, Adrian (2005), 'Lines of Flight + Art + Politics', in A. Parr (ed.), *The Deleuze Dictionary*, 146–8, Edinburgh: Edinburgh University Press.
Patterson, Angus (2012), 'The Perfect Marriage of Art and Industry', *Journal of the Antique Metalwork Society*, 20: 55–75.
Patton, Paul (1994), 'Translator's Preface', in Gilles Deleuze, *Difference and Repetition*, xi–xiii, New York: Columbia University Press.
Pehnt, Wolfgang (1983), *Das Ende der Zuversicht. Architektur in diesem Jahrnhundert. Ideen–Bauten–Dokumente*, Berlin: Siedler Verlag.
Perez-Gomez, Alberto (2012), 'The Historical Context of Architectural Representation', in Phil Ayres (ed.), *Persistent Modelling: Extending the Role of Architectural Representation*, 13–25, London: Routledge.
Peters, John Durham (2015), *The Marvelous Clouds: Toward a Philosophy of Elemental Media*, Chicago: University of Chicago Press.
Petreca, Bruna (2016), 'An Understanding of Embodied Textile Selection Processes and a Toolkit to Support Them', PhD thesis, Royal College of Art, London.
Pevsner, Niklaus (2010), *Visual Planning and the Picturesque*, Los Angeles: Getty Research Institute.
Pinson, Stephen (2012), *Speculating Daguerre: Art and Enterprise in the Work of L. J. M. Daguerre*, Chicago: University of Chicago Press.
Pirsig, Robert M. (1974), *Zen and the Art of Motorcycle Maintenance*, New York: William Morrow and Company.
Plant, Sadie (1997), *Zeros + Ones: Digital Women and the New Technoculture*, London: Fourth Estate.
Punter, David (2007), 'The Uncanny', in C. Spooner and E. McEvoy (eds), *The Routledge Companion to Gothic*, 129–36, London: Routledge.
Pye, David (1950), *Ships*, Harmondsworth: Penguin.
Pye, David ([1964] 1972), *The Nature of Design*, London: Studio Vista.
Pye, David (1978), *The Nature and Aesthetics of Design*, London: Barrie and Jenkins.
Pye, David ([1968] 1985), *The Nature and Art of Workmanship*, Cambridge: Cambridge University Press.
Pye, David (1986), 'Notes on Technique: Bowls and Boxes', in *David Pye: Wood Carver and Turner*, 43–52, London: Crafts Council.
Ratcliffe, Matthew (2018), 'Perception, Exploration and the Primacy of Touch', in A. Newen, L. De Bruin and S. Gallagher (eds), *The Oxford Handbook of 4E Cognition*, 281–99, Oxford: Oxford University Press.
Rees-Roberts, Nick (2017), 'The Spectacle of Process: Documenting Fashion on Film', in E. Mora and M. Pedroni (eds), *Fashion Tales*, 163–79, Berne: Peter Lang.
Ridyard, Simone (2015), *Archisketcher*, Apple Press.
Robbins, Christa (2002), 'Surface', *The Chicago School of Media Theory* [webpage]. https://lucian.uchicago.edu/blogs/mediatheory/keywords/surface/ (accessed 11 February 2015).
Rosen, Philip (ed.) (1986), *Narrative, Apparatus, Ideology*, New York: Columbia University Press.
Ross, Andrew (ed.) (1997), *No Sweat: Fashion, Free Trade, and the Rights of Garment Workers*, London: Verso.
Rossiter, Ned (2016), *Software, Infrastructure, Labor: A Media Theory of Logistical Nightmares*, New York; London: Routledge.

Rowe, Colin and Robert Slutzky (1963), 'Transparency: Literal and Phenomenal', *Perspecta*: 45–54.
Ruskin, John (1887), *Modern Painters*, volume 1, New York: Wiley.
Ruskin, John ([1851–3] 2009), *The Stones of Venice*. http://www.gutenberg.org/files/30755/30755-h/30755-h.htm#page151 (accessed 1 February 2019).
Sasaki, Atsushi (2017), 'Apichatpong Weerasethakul–Fever Room: Interview', *The Japan Foundation Asia Center*, 5 July. https://jfac.jp/en/culture/features/f-ah-tpam-apichatpong-weerasethakul/ (accessed 17 February 2019).
Scarry, Elaine (1998), 'On Beauty and Being Just', *The Tanner Lectures on Human Values*, 25–6 March. https://tannerlectures.utah.edu/_documents/a-to-z/s/scarry00.pdf (accessed 4 February 2019).
Scheerbart, Paul (1914), 'CXI: Die Glaskultur', *Glasarchitektur*. http://gutenberg.spiegel.de/buch/glasarchitektur–1752/3 (accessed 14 July 2016).
Scheerbart, Paul ([1913] 2012), *Lesabéndio: An Asteroid Novel*, Cambridge, MA: Wakefield.
Scheerbart, Paul ([1914] 2014), '*Glass Architecture*, 1914', in J. McElheny and C. Burgin (eds), *Glass! Love!! Perpetual Motion!!! A Paul Scheerbart Reader*, 20–91, Chicago: University of Chicago Press.
Scheiffele, Walter (1994), *Wilhelm Wagenfeld und die moderne Glasindustrie: Eine Geschichte der deutschen Glasgestaltung von Bruno Mauder, Richard Süßmuth, Heinrich Fuchs und Wilhelm Wagenfeld bis Heinrich Löffelhardt*, Stuttgart: Verlag Gerd Hatje.
Schirren, Mattias (2004), *Bruno Taut: Alpine Architecture, A Utopia/ Eine Utopie*, Munich and New York: Prestel.
Schulze, Franz (1985), *Mies van de Rohe: A Critical Biography*, Chicago: University of Chicago Press.
Schwanhäusser, Anja (ed.) (2016), *Sensing the City*, Basel: Birkhäuser.
Schwartz, Frederic J. (2009), 'Wilhelm Wagenfeld and Carl Jakob Jucker Table Lamp. 1923–24', in *Bauhaus 1919–1933: Workshops for Modernity*, 138–45, Museum of Modern Art.
Schwenger, Peter (2006), *The Tears of Things: Melancholy and Physical Objects*, Minneapolis, MN: University of Minnesota Press.
Sellafield Limited (2017), 'Army Manoeuvres', *Energise*, 101 (August).
Sennett, Richard (1992), *The Conscience of the Eye*, New York: W. W. Norton & Company.
Sennett, Richard (2008), *The Craftsman*, London: Penguin.
Sharp, Dennis (1972), *Glass Architecture by Paul Scheerbart and Alpine Architecture by Bruno Taut*, New York: Praeger.
Simondon, Gilbert ([1958] 1992), 'The Genesis of the Individual', in J. Crary and S. Kwinter (eds), *Incorporations*, 297–319, New York: Zone.
Simondon, Gilbert ([1958] 2017), *On the Mode of Existence of Technical Objects*, Minneapolis, MN: Univocal.
Sloterdijk, Peter (2013), *In the World Interior of Capital: Towards a Philosophical Theory of Globalization* (Wieland Hoban translation), Cambridge; Malden, MA: Polity.
Smelik, Anneke (2015), 'Fashioning the Fold: Multiple Becomings', in R. Braidotti and R. Dolphjin (eds), *This Deleuzian Century*, 37–55, Amsterdam: Brill Rodopi.
Smisko, Molly (2016), 'Assimilation', *Vimeo*. https://vimeo.com/176478169 (accessed 11 February 2019).
Smisko Ackerman (2018), 'Membrane', *Vimeo*. https://vimeo.com/266024562 (accessed 5 March 2019).

Sobchack, Vivian (2016), 'From Screen-scape to Screen-sphere: A Meditation in *Medias Res*', in D. Chateau and J. Moure (eds), *Screens: From Materiality to Spectatorship*, 157–75, Amsterdam: Amsterdam University Press.
Sole, Michael (1994), 'Electroforming: Methods, Materials and Merchandise', *Journal of Minerals, Metals and Materials Society*, 46 (6): 29–35.
Somaini, Antonio (2016), 'Walter Benjamin's Media Theory', *Grey Room*, 62 (Winter): 6–41.
Spivak, Gayatri (2012), 'Imperative to Re-imagine the Planet', in *An Aesthetic Education in the Era of Globalization*, 335–50, Cambridge, MA: Harvard University Press.
Sterling, Bruce (2011), 'An Essay on the New Aesthetic', *Wired*, 2 April. https://www.wired.com/2012/04/an-essay-on-the-new-aesthetic/ (accessed 5 June 2019).
Steyerl, Hito (2017), *Duty Free Art*, London: Verso.
Stiegler, Bernard (1998), *Technics and Time, volume* 1, Stanford: Stanford University Press.
Strauven, Wanda (2012), 'Early Cinema's Touch(able) Screens: From Uncle Josh to Ali Barbouyou', *NECSUS*, 1 (2): 155–76.
Stuart, John A. (1999), 'Unweaving Narrative Fabric: Bruno Taut, Walter Benjamin, and Paul Scheerbart's *The Gray Cloth*', *Journal of Architectural Education*, 53 (2): 61–73.
Stuart, John A. (ed.) (2001), *The Gray Cloth. Paul Scheerbart's Novel on Glass Architecture*, Cambridge, MA: MIT Press.
Taussig, Michael (2011), *I Swear I Saw This: Drawings in Fieldwork Notebooks Namely My Own*, Chicago: University of Chicago Press.
Taut, Bruno (2014), 'Glass House: Cologne Werkbund Exhibition, 1914', in J. McElheny and C. Burgin (eds), *Glass! Love!! Perpetual Motion!!! A Paul Scheerbart Reader*, 98–110, Chicago: University of Chicago Press.
Thiel, Philip (1996), *People, Paths, and Purposes: Notations for a Participatory Envirotecture*, Seattle, WA: University of Washington Press.
Thompson, Edward Palmer (1967), 'Time, Work-discipline, and Industrial Capitalism', *Past and Present*, 38: 56–97.
Tóth, Edit (2015), 'Capturing Modernity Jazz, Film, and Moholy-Nagy's *Light Prop for an Electric Stage*', *Modernism/Modernity*, 22 (1): 23–55.
Tschumi, Bernard (1994), *The Manhattan Transcripts*, London: Academy Editions.
Tschumi, Bernard (2014), *Notations: Diagrams and Sequences*, London: Artifice.
Tsivian, Yuri (2013), *Early Cinema in Russia and Its Cultural Reception*, London: Routledge.
Uhlirova, Marketa (2013), '100 Years of the Fashion Film', *Fashion Theory*, 17 (2): 137–57.
Ursprung, Philip (ed.) (2002), *Herzog & de Meuron: Natural History*, Montreal: Canadian Centre for Architecture.
V&A (2016), '"Cinderella" Table, by Jeroen Verhoeven, 2005–2006', *V&A* [webpage]. http://www.vam.ac.uk/content/articles/c/cinderella-table/ (accessed 27 June 2017).
Väliaho, Pasi (2010), *Mapping the Moving Image: Gesture, Thought and Cinema Circa 1900*, Amsterdam: Amsterdam University Press.
Vasseleu, Cathryn (2015), 'Simulated Translucency', in S. Cubbit (ed.), *Digital Light*, 163–78, London: Open Humanities Press.
Vermeulen, Timotheus and Robin van den Akker (2010), 'Notes on Metamodernism', *Journal of Aesthetics & Culture*, 2 (1): 56–77.
Vidler, Anthony (1992), *The Architectural Uncanny: Essays in the Modern Unhomely*, Cambridge, MA: MIT Press.
Viola, Bill (1997), '*The Veiling*', in *Bill Viola* [exhibition catalogue], New York: Whitney Museum of American Art.

Vismann, Cornelia (2008), *Files: Law, Media and Technology*, Stanford: Stanford University Press.

von Hippel, Frank, Masafumi Takubo and Jungmin Kang (2019), *Plutonium: How Nuclear Power's Dream Fuel Became a Nightmare*, Singapore: Springer.

von Schnitzler, Lilly (1929), 'Weltausstellung Barcelona', in *Europäische Revue* (April–September) Heft 1–6, V. Jahrgang, 1. Halbband, hrsg. von Karl Anton Prinz Rohan.

Waddingham, Anne (2014), *New Hart's Rules*, 2nd edn, Oxford: Oxford University Press.

Wagenfeld, Wilhelm (1963), lecture given in conjunction with the exhibition 'Deutsches Glas', Ladesgewerbeamt Stuttgart, 16. 12.1963, 2–3, courtesy of the Wilhelm–Wagenfeld Stiftung, Bremen.

Ward, Janet (2001), *Weimar Surfaces: Urban Visual Culture in 1920s Germany*, Berkeley, CA: University of California Press.

Weiner, Norbert (1948), *Cybernetics: or Control and Communication in the Animal and the Machine*, Cambridge, MA: MIT Press.

Welch, Celina R. (1993), 'Mies van de Rohe's Compromise with the Nazis', *Wiss. Z. Hochsch. Archit. Bauwes.—A.—Weimar*, 39 (1/2): 103–9.

Whyte, Iain Boyd (ed.) (1984), *The Crystal Chain Letters: Architectural Fantasies by Bruno Taut and His Circle*, Cambridge, MA: MIT Press.

Williams, Robert and Bryan M. Wilson (2013), *Cumbrian Alchemy*, Cumbria: University of Cumbria Press.

Wilson, Bryan M. (2013), 'The Nuclear Present', in E. Ellsworth and J. Kruse (eds), *Making the Geologic Now*, 222–5, New York: Punctum.

Woolgar, Christopher (2008), 'Wellington, His Papers and the Nineteenth-century Revolution in Communication' (University of Southampton, 20 November) [lecture]. https://eprints.soton.ac.uk/66250/1/CWInaugweb.pdf (accessed 30 July 2018).

Wumkes, Mark (2017–18), email correspondence with Benedict Carpenter van Barthold.

Yampolsky, Mikhail (1993), 'Transparency Painting: From Myth to Theater', in A. Efimova and L. Manovich (eds), *Tekstura: Russian Essays on Visual Culture*, 127–51, Chicago: University of Chicago Press.

Yerkes, Carolyn (2017), *Drawing After Architecture: Renaissance Architectural Drawings and their Reception*, Venice: Marsilio.

Young, Liam (2015a), 'An Atlas of Fiducial Architecture: A Tour through the Shadows of the Digital', *after us*. https://medium.com/after-us/an-atlas-of-fiducial-architecture-f2b5bb421ef8 (accessed 16 July 2018).

Young, Liam (2015b), 'New City: Machines of Post Human Production', *Centre de Cultura Contemporània de Barcelona*, 16 December. www.cccb.org/en/multimedia/videos/liam-young-talks-about-new-city-machines-of-post-human-production/222592# (accessed 16 July 2018).

Young, Sarah (2019), 'ASMR Videos', *The Independent*, 3 April. https://www.independent.co.uk/life-style/asmr-meaning-definition-what-is-youtube-video-reddit-a8852826.html (accessed 29 June 2019).

Zeitguised (2018), 'Emancipath', *foam Studio* [webpage]. https://foam.studio/kvadratexploration (accessed 11 February 2019).

Zheng, Caroline Yan (2019), *Feu et Bois* [webpage]. https://feuetbois.net/ (accessed 30 July 2019).

Index

Page numbers in *italics* refer to figures and their captions, page numbers in **bold** refer to tables.